Raising Beef Cattle

An Essential Guide to Raising Cows, Calves, Bulls, Steers and Heifers in Your Backyard or on a Small Farm

Contents

INTRODUCTION 1

CHAPTER 1: THE 6 BENEFITS OF RAISING BEEF CATTLE 2

A DOZEN THINGS YOU DIDN'T KNOW ABOUT COWS/CATTLE 3

6 REASONS YOU SHOULD CONSIDER GROWING YOUR OWN BEEF CATTLE 9

CHAPTER 2: BEEF CATTLE BREEDS AND SELECTION 12

THE TOP 9 MOST COMMON BEEF CATTLE BREEDS IN THE UNITED
STATES 13

BEEF CATTLE SELECTION 16

PUREBRED VS CROSSBRED 19

CHAPTER 3: CATTLE PSYCHOLOGY AND HANDLING 22

VISION AND CATTLE BEHAVIOR 22

VISION AND CATTLE HANDLING 23

LIGHT AND CATTLE BEHAVIOR 24

LIGHTING AND CATTLE HANDLING 25

NOISE AND CATTLE BEHAVIOR 25

NOISE AND CATTLE HANDLING 26

TOUCH AND CATTLE BEHAVIOR 27

HEALTH AND CATTLE BEHAVIOR 28

THE FLIGHT ZONE 29

CHAPTER 4: FACILITIES, HOUSING AND FENCING...............................31

BEEF CATTLE HANDLING FACILITIES 32

BEEF CATTLE FEEDING EQUIPMENT 34

BEEF CATTLE HOUSING 36

CHAPTER 5: BEEF CATTLE NUTRITION AND FEEDING.......................40

THE RUMINANT DIGESTIVE SYSTEM 42

NUTRITIONAL REQUIREMENTS OF CATTLE...................................... 44

TYPES OF CATTLE FEEDS... 46

CHAPTER 6: YOU CAN STILL MILK YOUR BEEF COWS! 49

BEEF VS. DAIRY CATTLE.. 49

MILK PRODUCTION IN BEEF CATTLE 51

HOW TO CHOOSE THE BEST BEEF COWS FOR MILK PRODUCTION 52

DUAL-PURPOSE CATTLE... 53

HOW TO MILK A COW CORRECTLY 54

CHAPTER 7: BEEF CATTLE HYGIENE, HEALTH, AND
MAINTENANCE 56

BEEF CATTLE HYGIENE... 57

BEEF CATTLE HEALTH 61

COMMON CATTLE DISEASES 61

COMMON CATTLE PARASITES 64

FOOD POISONING 65

UREA POISONING.. 66

CYANIDE AND NITRATE POISONING 66

VACCINATION 67

CHAPTER 8: BULLS AND STEERS..69

BULLS VS. STEERS 70

CHAPTER 9: COWS AND HEIFERS 76

ANATOMICAL DIFFERENCES BETWEEN COWS AND HEIFERS............. 76

BENEFITS OF RAISING REPLACEMENT HEIFERS................................ 85

BENEFITS OF BUYING REPLACEMENT HEIFERS 85

CHAPTER 10: CATTLE BREEDING AND REPRODUCTION 87

HOW TO PREPARE YOUR COW TO CONCEIVE SUCCESSFULLY 88

CARING FOR OLDER COWS 92

CHOOSING THE RIGHT BULL FOR BREEDING 94

HOW TO CARE FOR YOUR BULL 95

CHAPTER 11: CALVING AND CARING FOR NEWBORNS.........................99

FEEDING YOUR PREGNANT COWS AND HEIFERS 100

OTHER THINGS TO PAY ATTENTION TO 101

SIGNS OF CALVING 102

PREPARING FOR CALVING 103

CALVING 103

CALF HANDLING 104

CALF NURSING 104

CALF HEALTH 105

CALF IDENTIFICATION 105

CALF CASTRATION AND IMPLANTS 106

CHAPTER 12: EXPERT TIPS FOR YOUR BEEF CATTLE BUSINESS 107

A QUICK GUIDE FOR BEGINNERS 107

PEOPLE YOU NEED TO KNOW 110

21 TIPS FOR RUNNING A COMMERCIALLY SUCCESSFUL CATTLE REARING OUTFIT 110

CONCLUSION 115

HERE'S ANOTHER BOOK BY DION ROSSER THAT YOU MIGHT LIKE 116

REFERENCES 117

Introduction

Americans love beef - so much so we consume 25 billion pounds of it each year. But there's something far more interesting and fulfilling than eating the juiciest medium-rare steak on a Friday night, and that is raising your own beef.

The satisfaction of knowing that the beef you eat is obtained from cattle raised in a safe, healthy, cruelty-free environment is better experienced than explained. You probably agree, and that's why you have this book in your hands.

In the pages that follow, you will gain in-depth knowledge on topics ranging from breed selection to cattle psychology to housing, nutrition, reproduction and calving. Smart tips for your beef cattle business are also included in this book to help make sure your new venture is as financially rewarding as it is satisfying.

Are you ready to take a leap into your new adventure? Then flip the page and let's begin!

Chapter 1: The 6 Benefits of Raising Beef Cattle

Did you know that every 14th day of July is Cow Appreciation Day? No, it's not a national holiday, but Chick-fil-A did launch this event in 2004 as a way to get people to spare a thought for the cows that provide us so much.

On Cow Appreciation Day, people are encouraged to hug cows, thank dairy farmers in person, or buy locally made milk and dairy products. On the fun side, certain Chick-fil-A branches also offer a free meal on this day to any customer that comes into their restaurant dressed as a cow. Sweet, right? And it's the least we can do considering how much cows do for us, often at the cost of their lives.

Over 98% of a beef animal will be used when processed. Beef, cheese, milk, ice cream and yogurt are delicious consumables that come to mind when we think of cows (or cattle more correctly. Not every animal that looks like a cow is a cow. But more on that later). From cattle, we get about 25 billion pounds of beef annually. That's huge!

But there's so much more that cattle do for us. About 45% of a cow's body is used as meat and the other parts go into the production of glue, leather, soap, gelatins, pharmaceuticals, china and even insulin. Picture this: one bovine animal provides enough hide to make about 144 baseballs, or 20 footballs or 12 basketballs.

Cattle are also wonderful recyclers as they feed on many of the by-products from manufacturing products like potato chips, candy, and beer. We haven't even mentioned the amazing benefits their dung provides in manure. One bovine can produce up to 80 pounds of manure every single day! That's more manure than even scientists or farmers know what to do with.

There are many benefits we enjoy from rearing cattle, and so it's difficult to think about them beyond all the ways they are useful to us as humans. But on their own, cattle are fascinating animals worth looking closely at. Here are just a few fun facts to make you look at cattle a little differently.

A Dozen Things You Didn't Know About Cows/Cattle

1. Cattle Have Nearly 360-Degree Vision

This means it's practically impossible to sneak up on them. It also means that cow tipping is mostly wishful thinking. Besides their nearly panoramic vision, they also never zone out.

Plus, come on, these animals are massive! We are looking at about 1500 pounds average effectively balanced on four legs. What are the odds? Watching someone try to trip a sleeping cow might be funny if it wasn't so dangerous.

2. The Word "Cattle" has its Roots in the Word "Chattel"

Yes, the same Anglo-French term that means "personal property." Back in the day, cattle were considered valuable property and a person's wealth was measured by it.

3. Every Cow is Female

Every cow is a girl or more correctly, a female that has birthed a calf. Those that haven't are called heifers. Males, on the other hand, are called bulls. If they are castrated so they can no longer breed and are reared for their beef alone, then they are called steers.

There are also those called veal. These are specifically raised to reach a maximum weight of about 500 pounds.

Other names used to differentiate members of a herd include:

Stag: A stag is just like a steer except that it is also used as a "gomer bull." Gomer bulls are used to detect heifers and cows in heat.

Ox: These are raised specifically to do draft work. Draft work includes pulling farm and travel machinery such as wagons, plows, or carts.

Oxen are mostly castrated male bovines, but they can also sometimes be bulls or even female cattle.

Freemartin: Freemartins are infertile heifers. Infertility in heifers is usually the result of sharing the womb with a bull calf. The testosterone levels produced by the bull calf in the womb affect the production of estrogen in the female calf.

Freemartins can be born in one of two ways. They can have underdeveloped reproductive organs, or they might have both male and female parts (such freemartins can also be called hermaphrodites).

Hermaphroditic freemartins will usually develop secondary male characteristics as they mature, such as a wide forehead or a muscular crest around their neck.

Cattle: A plural term used when there's more than one bovine, especially when the genders are mixed or uncertain.

4. Cows are not always Black and White

Bulls are not always solidly colored either. With cattle, color is determined by breed and not by sex. Cattle can literally come in many colors and these are varied by different markings. You can find cattle in brown, yellow, white, black, red, gray, and even orange. They can also come in a variation of these colors (most times mixed with white) such as speckled, pointed, patchy, dorsal-striped, white-faced or white-tailed.

Generally...

Friesians, Holstein-Friesians and Purebred Holsteins, male and female, are always black and white.

Dairy cows such as Jersey, Guernsey and Brown Swiss cows are usually either solid red or solid brown.

Beef cows such as the Limousin, Gelbvieh, Red Brangus, Red Angus, Simmental and Santa Gertrudis breeds are also usually red or brown.

Belgian Blue cattle, though, are not actually blue. They are more bluish roan than blue. They only appear smoky-blue because of the way the white and black hairs on their coats are mixed.

5. Both Sexes of Cattle can either be Horned or Polled or Both

So, it's not a great idea to rely on the presence or absence of horns or polls in determining the gender of bovine. To tell the gender of a bovine accurately, look behind the animal's hind limbs to see if there's a scrotum or an udder.

6. Bulls Cannot See Red

Like their other bovine brothers and sisters, bulls are red/green colorblind. So, why do they charge at the matador then? It turns out all flapping in the breeze that aggravates them, which is understandable, right?

So, even if the matador used an indigo flag, the bull would still charge. But why have they continued to use red flags? Well, it isn't ignorance. It's actually a more "sinister" reason - to hide the bull's blood.

7. Named Cows Produce More Milk than Nameless Ones

This suggests that the more emotionally invested a farmer is in their relationship with their cows, the more milk the cows produce. A good relationship with humans means that cows are less stressed when milked, which means more milk for the farmer. A cow called by name will produce nearly 500 more pints of milk in a year!

But if the cow feels jittery around their human, she gets stressed, and her body produces cortisol. This hormone inhibits the production of milk, which means less milk. But even more important, a happy loved cow is less likely to hurt her human when being milked.

8. Cows are very Social and even have Best Friends

It's hard for you to find cows all alone except when they are ill or about to give birth. Separating a cow from her best friends could cause her to become stressed. The body secretes more cortisol (the stress hormone), and heart rate goes up when you put them with random bovines rather than their preferred partners.

9. Cows can Swim

Cows are excellent swimmers, believe it or not. After Hurricane Dorian, three cows were found at the Cape Lookout National Seashore. They were believed to have swum all the way from Cedar Island, where they lived before the hurricane-ravaged their home.

Swimming about 4 to 5 miles might seem astonishing, but experienced farmers won't find this information surprising.

10. Surrogacy is a thing Among Cows

There are such things as surrogate cows. Surrogacy is becoming more common these days, especially with dairy cows. The process involves moving embryos from genetically superior cows to other less superior cows.

Naturally, cows produce only one embryo at a time, but when surrogacy is the plan, the cow gets injected with a hormone triggering the production of many eggs, which are then fertilized.

The fertilized eggs (embryos) can be as many as 80 or even 90 but, in the end, only about 6 to 7 end up being usable.

The vet removes the embryos from the cow, employing a process called an embryo flush. These embryos can then be transferred to the less superior surrogate cows so their offspring are of better genetic quality compared to what they might have produced on their own.

Surrogacy among cows is a genius innovation, and not just because you get to "create" your preferred, desirable, genetically superior cows. Through surrogacy, farmers in other countries with inadequate resources to meet dairy-cow demands can improve their own bovine gene pool, producing quality cows to meet their lack.

11. Aurochs Are Cattle's Earliest Ancestors

Aurochs were huge, wild beasts originally located in the Indian subcontinent before they spread to China, the Middle East, North Africa and then Europe. After a while, about 8,000 to 10,000 years ago, people domesticated aurochs.

In 1493, Columbus introduced these domesticated aurochs, now known as cattle, to the western hemisphere. Later, in 1519, Hernando Cortez, a Spanish explorer, took the offspring of these cattle to Mexico. In 1773, Juan Bautista de Anza supplied the early California missions with 200 head of cattle. And this was how cattle evolved and spread around the world.

Cattle are raised and bred globally, in widely varied settings and climates. This is possible because cows can survive and even thrive eating only low-quality grasses and feed. For most cattle, grazing would be on steep, hilly, rocky or dry grounds, unsuitable for cultivating crops or building houses.

12. Cattle Have 32 Teeth in Total but Have no Upper Front Teeth

So, how do they cut grass? Well, to cut grass, they join their lower front teeth to the hard upper palate. After doing the cutting, they then chew their food roughly 50 times in 60 seconds. This means they move their jaws close to 40,000 times per day!

It's common knowledge that cattle have four-compartment stomachs, the rumen being the main stomach. The rumen is the part of the stomach that holds the partially digested food called the cud. From this part of the anatomy the cud goes back into the cattle's mouth when regurgitated.

The rumen can hold about 50 gallons of partially digested food. If you want to feel small, the human stomach can only hold 1 quart of food, which is just about a quarter of a gallon!

A bovine typically spends a third of its day eating, consuming about 40 pounds worth of food and about 30 to 50 gallons of water each day. Naturally, cattle pass out large amounts of dung and urine daily. It is estimated that cattle produce about 60 pounds of manure and roughly 30 gallons of urine every day. In one year, we are looking at over 20,000 pounds (or 10 tons!) of manure!

This makes an excellent point to segue into the next part of this chapter. Let's see all the reasons you should raise your own beef cattle.

6 Reasons You Should Consider Growing Your Own Beef Cattle

1. Raising Your Own Beef Cattle Can Help Improve Your Land

When done properly, grazing your cattle can help improve the quality of your land. Here's how that works.

Plants need leaves to photosynthesize. When they do that, they release sugars into the soil, which soil microbes then use to break down soil nutrients, making these nutrients available for the plants to use. By grazing on the grass, cattle break up the ground and remove the old foliage so that the grasses have room to germinate and photosynthesize.

Also, by indirectly laying plant debris on top of the soil, cattle assist in the maximizing of the life cycle of soil minerals. As the debris composts, it makes it possible for these nutrients to properly cycle and enter the root nodules of plants.

Plus, don't forget their urine and manure, which supply nitrogen, additional microorganisms, and more partly decomposed grass to the soil. Thanks to this natural fertilizer, your soil will be infiltrated and hold water better for future plants. This means that your land will be less vulnerable to drought.

Another very important thing your grazing cattle do as they improve your land is to sequester atmospheric carbon in large amounts. This creates a safer ecosystem for us all.

2. You Get Access to Healthy Meat

From a single cow, you can get over 500 pounds of beef. Even if you eat that in one year, you'd still get to eat 1.3 pounds of beef each day, and that's substantial. Killing one cow to get that much beef is much more efficient, plus more humane, than killing the required number of chickens it would take to get the same meat.

Also, raising your own beef cattle, you get access to the choicest, healthiest parts. There's the liver, for instance, which is the most nutritious food you can find. You should eat about 1 to 3 ounces of liver per week, at least. Other parts like shanks, brisket, and oxtail, when made into bone broths, are also super healthy for you as they provide rare nutrients you can't get from other sources.

Furthermore, beef from grass-fed cattle is richer in healthy fatty acids, which are important for proper immunity, heart and brain function. Research shows that the fat ratio you find in grass-fed beef is very similar to the ratio found in the ancestral human diet.

Pasture-raised cattle also contain the highest levels of conjugated linoleic acid, which has been known to have anti-cancer properties. And compared to grain-fed beef cattle, they supply 7 times more beta-carotene, 2 times more vitamin B2, and thrice as much vitamin B1.

3. You Get More Nutritious Crops

In recent years, nutrient density in vegetable crops in America has witnessed a steady decline. As of now, nutrient density in crops has dropped by up to 40% because livestock is nowhere to be found on our farmlands anymore.

Removing livestock from the land means you no longer get the benefits explained in the first point. Therefore, there would be a decrease in the available soil nutrients for plants, which will, in turn, affect the quality of the crops you harvest.

4. It's a Blossoming Income Stream

Pasture-raised cattle beef is not mainstream or even popular yet. Of the 30 million cattle we see on the market annually, just 1% are grass-fed. But from all indications, the market for grass-fed beef is growing, as consumers are becoming more aware of the positive benefits. Plus, grass-fed beef tastes sublime when people use the best methods to produce them, and the demand for these flavorful delicacies is on the rise.

To correctly breed your cattle, it's important to feed them in the fields or in the feedlot with high-energy feeds. This is the only way to make them fat and delicious. Feeding them on lush, delicious high-protein green grass will affect the flavor of your beef. It might seem counterintuitive, but it's the way it works. High-energy, carbohydrate feeds are much better at preparing cattle for processing than high-protein feeds.

High-protein will give your cattle stronger frames and better yield, but the luscious fat that gives beef its great taste, that's the work of carb diets. This is the key to producing the "best-beef-you-ever-tasted" kind of grass-fed beef.

5. Carefully Planned Grazing Increases Biodiversity

If cattle are properly grazed, their grazing activity can increase the biodiversity of the pasture in which they feed. This improves the ecosystem for millions of other critters found in that biome.

6. It's a Fun, Educational Project for the Whole Family

Nurturing cattle involve activities that can be carried out by different family members of different ages and skill levels. Raising cattle as a family can foster family bonding. And then there's all the fun of traveling around the country exhibiting your animals in cattle shows.

But showing cows isn't the only way raising cattle can be great for your family. The venture of raising cattle all on its own is rewarding. Everyone gets to learn dependability and develop a great work ethic.

Cattle are demanding animals, requiring care every single day in the sun and in the rain. Together, your family can work out the kinks of caring for them by drawing up budgets, making purchasing decisions, assigning responsibilities and managing money and other resources.

It's inevitable that doing all these things together as a family will bring you all closer and increase your love and appreciation for other living creatures with whom you share this planet.

Chapter 2: Beef Cattle Breeds and Selection

Beef cattle are specifically bred for their meat because of how efficiently they convert feed into meat. When feeding, they absorb the minimum amount of nutrients they need to carry out basic physiological functions. After that, they begin to gain weight, which is mostly muscle and not fat or bone. Thanks to this genetic predisposition, a newborn 90-pound calf only needs as little as 12 to 13 months to achieve butchering size.

Beef cows do produce milk - just not in large amounts. They produce just enough to keep them in tip-top physical shape as they rear their young calves.

That said, efficiency in growth isn't the only reason they are reared. There are other important qualities for which such herds are raised. These attributes include reproductive superiority, efficiency in the feed, and hardiness, which are critical traits when raising and breeding the best with minimal care and maintenance.

Genetics aside, they are also identifiable by their looks. Good quality herds look rectangular with broad chests, wide shoulders, a thickness along the top of their backs, and round and full stomachs and ribs. Good beef herds are never bony or skinny. They always look plump and robust but such animals typically cost more than the skinnier ones. So, remember this when shopping.

When it comes to color, they almost always come in solid colors. This can range from solid black to white to gray to red. It's rare for them to be spotted, although they do exist. Nonetheless, because more buyers prefer solid black, even preferring to pay a premium price for them, spotted cattle are becoming rare.

The first step in any venture is to select a breed. We've talked a bit about the breed in general, but these guys come in many breeds. Choosing the right one that matches your goals and objectives as a farmer is critical to enjoying a more profitable herd-growing experience.

Now, pay close attention because, in this chapter, you will be getting key pointers for picking the best breed for your needs. But, before that, let's look at the most common beef in the United States.

The Top 9 Most Common Beef Cattle Breeds in the United States

1. Black Angus

The Black Angus is the most popular breed in America. There are more than 330,000 of these animals currently registered, and the reason for its popularity is the value of its carcass. It is commonly touted knowledge that the cadaver of the Black Angus is well-marbled and very flavorful. They don't need a high level of maintenance, especially during the calving season. They are very efficient with feed and are excellent mothers too.

2. Charolais

There are many who believe that introducing the Charolais revolutionized the North American beef industry. Before the breed was introduced, American farmers were in search of heavier, larger-framed cattle, something they were not getting with traditional British breeds. But with introducing the Charolais, that problem was solved immediately.

Charolais are usually creamy white, or white. In the summer, their hair is short, and as the weather gets colder, the hair thickens to protect them.

3. Hereford

Herefords are desired for their fattening ability and early maturation. They are usually reddish yellow to dark red and have a white face. They are also quite docile, great mothers, good milkers, and typically live longer.

4. Simmental

The Simmentals are an old breed, widely distributed across the globe. They are usually white and red. They first entered the United States in the 19th century and have been a part of the American beef community since then.

Simmentals have a large frame with an impressive ability to gain weight.

5. Red Angus

The Red Angus is a less popular breed than its cousin, the Black Angus, but both breeds share the same favorable characteristics, like amazing marbling and terrific flavor. These cattle are docile, excellent mothers, and can tolerate hotter climates better than the other breeds accustomed to Highland conditions in Europe.

6. Texas Longhorns

Texas Longhorns are white and red with characteristically long horns. They have great calving ability and hybrid vigor when crossed with other varieties.

Texas Longhorns beef is choice meat because it's lean and low in fat, cholesterol and calories when compared to other kinds.

7. Gelbvieh

The Gelbvieh is European, but it got introduced to the US via artificial insemination. The breed is typically red and horned, although there are polled varieties that came about from crossing with hornless female cattle.

Many of the breed's best attributes include their great fertility, ease of calving, good mothering ability and impressive growth rate for calves.

8. Limousin

The Limousins are golden red and are mostly found in south-central France — Marche and Limousin, to be precise. The carcass merit of this breed is top notch, making it another popular beef in the industry.

9. Highlands

Highlands are known for their double coat and longhorns. These herds are super easy to keep as they often go by with the minimum in terms of feed, shelter and the likes. They do nicely in colder climates and are found thriving in Alaska and Scandinavian countries. They also succeed in southern climates like Georgia and Texas.

This breed is practically immune to eye infections and diseases such as pink eye and eye cancer, thanks to the forelocks and long lashes that protect their eyes.

Beef from Highland cattle is rich in flavor, well-marbled, and with only little waste fat.

Beef Cattle Selection

Selecting the right ones to raise will depend on your personal goals. Many people prefer to raise them to graze their pasture. Others want to sell feeder calves or raise them for showing. The most popular reason for keeping them is to get quality beef either to eat or for sale. So, you need to clarify your cattle-raising aspirations first before deciding on which breed to select.

Once you've done that, the following are a few factors you want to consider when selecting a breed.

1. Local Availability

It will be much easier for you if you get cattle common to your locale, except for when your heart is set on a particular type. So, check on neighboring farms in your vicinity and discover the breeds they raise when making your selection. There are several advantages to this.

First, if you go for those that are popular in your region, you'd have a larger pool to select from, which gives you more options. Plus, you won't have to spend big bucks moving them from one part of the country to the other. Considering the size of these animals, transportation costs can be large.

Also, because they are "natives" of your region, you won't need to pass them through the rigors of adapting to a new climate and new feed. Buying similar product to your neighbors' means you begin your enterprise with an already proven track record of success.

2. Hair Color

Hair color varies widely, as we discussed earlier, but following current trends, solid-colored cattle sell for prices much greater than the spotted varieties. While solid colored cattle are the priciest, they are most common in this industry due to higher demand. As with human fashion, black hides flaws like lack of muscle or fat and gives the bovine a more flattering look, hence their desirability.

Uniform color is more attractive to potential buyers. Many believe that a uniform herd grows, feeds, and attains butchering size at the same rate, even though this isn't necessarily always true. Whatever the case, achieving a uniform look is probably a good idea if you plan to sell, and much more easily achieved if all your cattle are a solid dark color.

There's a small catch, though. Black cattle do not stay cool as easily as lighter-colored cattle. So, offer plenty of water and adequate shade for them.

3. Horns

Generally, raising them without horns is often easier than those with horns. There's the danger factor — they can be sharp and tough! Then you also must consider all the space they require, both during transportation and at the feeding bunker.

For these reasons, horned cattle sell for a price considerably lower than their counterparts (polled cattle). But especially if you're working with limited space - perhaps just your backyard - you might have to spend the extra bucks and choose the hornless varieties.

Now, remember that once dehorned, they are different as hornless (polled) cattle. Dehorned are considered bovines who have had them physically removed. If you don't want horned offspring, then you shouldn't get dehorned cattle as they still have the genes to produce calves with horns and can pass this trait on. Polled cattle, on the other hand, are bovines born without horns. These do not have these genetics, and won't pass them down to their offspring.

Breeds like the Angus, Polled Hereford, and Polled Shorthorn are naturally always polled but, for other varieties, horning, while possible, should be avoided using selective reproduction.

Now, since we're here, it's important to talk about the dangers of dehorning. Dehorning causes a high level of stress to the animal, plus the potential complications. A much better and safer way to remove horns remains through selective breeding.

4. Breed Characteristics

Having looked at all other factors like horns, hair color and local availability, it's time to check out the characteristics of the cattle. Your knowledge of these features and how they are affected by environmental changes can be helpful to determine the best breeds you want to raise in your situation.

The following are the important qualities to consider thoughtfully:

Carcass Merit: The carcass of an animal includes all that's left after the hide, head, and internal organs have been removed. It typically consists of fat, bone and muscle.

Carcass merit refers to an assessment of the yield (also known as lean meat) and its eating quality. If you plan to sell your animals on the grid, this determines the price offered for the beef. The higher the value, the higher the prices you're offered, and the more satisfied customers you'd have.

Pick those that are identified for their lean meat. No one likes to drain grease from meat. Also, it is very important to look for their eating quality and tenderness.

Body Size: Specifically, this refers to adult body size. Large parents mean large calves, which usually means birthing difficulties most of the time. But large cattle mean you get calves with heavy weaning weights, which can be a great if you plan to sell feeders by weight.

But do remember that with a large size comes an even larger responsibility. These guys sure do eat! If you live in an arid climate with little grass, you probably shouldn't be thinking of getting a 1,500-pounder. It's not a good idea to get large cattle if you're working with a small space.

Milking Ability: This refers to just how much milk a cow can produce to feed her calf. The more milk, the heavier the weaning weight of her calves, but cows that milk heavily are often skinnier because their bodies direct all the calories toward milking. Therefore, these cows take longer to rebreed.

If you have a heavy milking cow, the quality and quantity of feed you give your cow really matters. These cows do not do well with a meager pasture.

Growth Rate: A measure of how much a bovine can grow during a period, as well as how much feed is required to produce one pound of weight gain. This rate is expressed using the unit, ADG, which stands for "average daily gain."

To calculate the average daily gain of a cattle animal, simply divide the pounds gained over a period by the number of days. An ADG of 3 or more is considered high.

Cattle with a high ADG, naturally need high-quality, high-energy feedstuff to achieve their fullest growth potential.

Adaptability: How well the breed thrives in challenging environmental conditions such as sparse feeds, or extreme weather, or in the presence of insects should be considered. This is the cattle's adaptability.

Although this factor has been mentioned last, it's important because what's the point of having a fantastic calf with a high ADG and carcass merit that dies because it can't cope with the challenges of the environment? Exactly.

Purebred vs Crossbred

Purebreds are cattle with parents from the same strain, while a crossbred will have each parent from different or unknown breeds. Both have their strengths and demerits. Let's look at each in detail.

Purebreds

With purebreds, you can be part of a recognized breed association like the American Angus Association, for example. The benefit of being part of such organizations is that they promote and help you raise your breed by providing the right education and supporting your marketing effort.

If you're going into the raising and selling of your reproductive stock, then purebreds are your best bet. You may come to where you want to sell crossbreeds too, but even at that, you'd still need your purebreds to serve as the foundation breeding lines. Also, shows, fairs and competitions are more open to purebreds than to others.

Looking at the marketing angle, it's a broader space for purebreds. For instance, you can only find registered trademarks (like Certified Angus Beef, for example) for purebreds and not for crossbred animals.

Now, if you do go for purebred cattle, you must be on top of all things that are data and paperwork-related. You need to do this because it's the only way to register your animals.

Registration papers contain information on the animal, including its parentage, and expected breeding performance. Without registration, your cattle will not be eligible for shows, but when your animals are properly registered, you'll be offered greater prices for them as having a registered animal increases its resale value.

Crossbreeds

Crossbreeds have the advantage of heterosis or hybrid vigor. This refers to how they excel in key performance areas such as fertility, growth and longevity. They typically do a lot better than both their purebred parents in these areas. This often is to be expected, seeing as the main aim of crossbreeding is to obtain the best traits of two purebreds in one superior offspring.

Certain features have low heritability; they are not easily passed to the next generation. Examples of such attributes include mothering instinct, reproductive performance, and environmental adaptability. Crossbreeding can help to improve such features to increase their heritability.

Should you get crossbred cattle? Why not? Well, that depends on two things. If your cattle-raising goals involve purebreds and purebreds alone, you should stay away.

Now, if the aforementioned isn't your plan, then it is okay to choose crossbred cattle. They have are advantages over purebreds. Superior mothering instincts, excellent fertility, longevity, and calves with heavier weaning weights are things you want in a cow, and crossbred cattle provide these traits and skills better than purebreds.

With crossbred bulls the benefits aren't as clear-cut. Mating these cows with crossbred bulls does not always go as expected. Often, variations are too wide, and the calves obtained differ widely in size and weight, which you don't want. So, to improve certain traits in your herd using breeding selection, it might be better to use a purebred bull.

A quick tip if you're going for crossbred cattle. Like George Orwell said in <u>Animal Farm</u>, all crossbreeds are equal, *but some are more equal than others.* If you missed it, the point is that all forms aren't all created equal. So, when buying this type of cattle, confirm from the seller that they are an offspring of different purebred breeds, not just a calf with unknown ancestry.

Chapter 3: Cattle Psychology and Handling

While animal cruelty and abuse are usually a result of terrible motives on the side of the handler, sometimes a handler uses cruel actions because they are frustrated and have run out of ideas.

In this chapter, you'll be exposed to how cattle behave and why they behave the way they do. Once you understand normal cattle behavior, you can learn to handle them without having to resort to cruel and dangerous measures. Let's get started.

Vision and Cattle Behavior

Cattle eyesight differs significantly from that of humans. And this difference is probably most evident in the relationship between their eyesight and their movements.

Cattle have a wide angle vision that allows them to see things happening beside them. So, if your cattle notices movement from the corner of their eye, regardless of how subtle the movement is, it will most likely balk and stop moving. Worse still, the perceived movement can frighten them and cause agitation that you'd rather not have. Introducing a paddle at this point or forcing them to keep on moving can lead to very unpleasant circumstances.

Besides their wide-angled vision, cattle do not have a good depth perception at ground level. For them to figure out how deep a hole is, they must lower their heads, so they can see the ground.

So, if your cattle are walking and notice a dip or hole, like a drain or even a change in ground texture, they'll most likely stop moving. You might even notice a few of them checking out what is on the ground.

If your cattle suddenly stop walking, your first instinct shouldn't be to force them to keep moving but to discover why they've stopped moving.

However, this sudden stop can be prevented if you consider the suggestions made in the next section.

Vision and Cattle Handling

Let's start with paying attention to their wide-angle vision. Because cattle can perceive movement from the corner of their eyes, and that movement can prevent them from moving, so eliminate distractions.

So, in building your handling space and/or loading ramp, add in a few side slabs high enough to keep the distractions out.

Besides ensuring the free flow of movement, blocking out the distractions can contribute to making your cattle less agitated. Make sure that they can't see people and things they can't control, as this will help keep them calm.

Building solid ramps that will block out the distractions is especially important for new cattle breeders because, as a new cattle breeder, your cattle are not familiar with your farm and are not used to those distractions. They have also most likely not yet been trained to ignore the distractions.

Something else you should do as regards your cattle's wide-angle vision is to remove anything around them that moves.

So, there should be nothing hanging anywhere or flapping in the breeze. Coats, hangers, and even tree branches should be removed. If you pull off your coat because the weather is too hot, do not hang it on the fence or anywhere within your cattle's line of sight.

If you have a barn, consider other ventilating mediums besides fans because fan blades can be distracting for your cattle.

Now, let's look at their depth perception. Make certain that the path that your cattle will be walking is free of any obstacles. Your cattle must be able to sense that walking a particular path will not be dangerous for them.

Make sure there are no drain chutes along the path. Also, make sure that the ground texture has the same consistency; there should be no ridges or crevices. There should also not be any puddles of water, as these can be perceived as a potential drowning site.

If your cattle suddenly stop to check things out on the ground, allow them to satisfy their curiosity. They will be more willing to listen to you if they have confirmed that they are not in danger.

Light and Cattle Behavior

Cattle find it easy to move from a place that isn't properly lit to one that is, but they will not go toward a brightly lit place if it's too bright.

Also, cattle rarely take well to shadows, whether those are on the walls or on the floor. Shadows confuse them as the shadows make it difficult for them to see what's ahead of them.

Lighting and Cattle Handling

Make sure that the destination (most likely a loading ramp or trailer) you want your cattle to move onto is better lit than the loading chute (race). You can do this by beaming light directly on the loading ramp.

This light should not be too bright, as that can deter your cattle. And the light should also not be shined directly into your cattle's eyes, as that can make them agitated.

As for the chute, make sure that the entire chute is evenly lit. There should be no shadows or dark spots. These make your cattle feel uneasy. The idea is just to make sure that your cattle know that what they see is all that is there.

Noise and Cattle Behavior

Cattle do not like loud sounds. And probably unfortunately for them, they have good hearing. To put things in perspective, in the best-case scenario, humans hear at 3000 hz while cattle can hear at up to 8000 hz. So, it makes sense that noise easily irritates them.

However, the emphasis here is on loud because cattle have no issues with white noise and random radio talk, provided the sound is at a reasonable volume. In fact, consistent white noise can help your cattle relax.

Cattle are disturbed by loud and sudden noises. The sounds from bells, trains, heavy-duty trucks, firecrackers and even sounds of heavy-duty machinery at a slaughterhouse is auditory sensory overload for your cattle.

Something else that they do not take well to shouting. So, whether it's being shouted at or just generally having humans screaming and shouting around them, you want to keep your vocalizations down when you're in their presence.

Now, cattle that generally have a calm disposition might not seem agitated at the sound of something loud and foreign, but this does not mean they aren't. Cattle that are calm will usually tilt their ears in the general area of foreign noise, as if they are trying to figure the sounds out.

Now that you know all that about how noise can affect your cattle's behavior, what can you do about it?

Noise and Cattle Handling

First, if you are running a slaughterhouse, make sure that the barn or wherever your cattle are housed is far from the slaughterhouse; far enough away that your cattle cannot hear the activities going on in there.

Also, if you love firecrackers, you must sacrifice for your cattle. Make sure that everybody knows that your property is a no-firecracker zone.

However, if you have a small farm, it might be difficult to cut out every source of noise completely. Here, having constant white noise play in the barn can help make your cattle less susceptible to reacting loud noise. You can use a radio to play talk show stations in your barn with the volume set to normal.

Make sure that you do not shout at your cattle. Not only will the shout not be effective, but it can also make the situation worse.

You also want to make sure that whatever disagreement you are having with anyone is away from your cattle's range of hearing, especially if you feel that the disagreement could lead to a shouting match.

And in rounding up your herd, whistling or shouting should not be considered, as those are classified as a sudden noise. But (and this might seem like an antithesis of all that has been said) blowing a horn to call your cattle could be a good idea. But you need to know if you want to use a horn, your cattle must be trained (with rewards) to come to the sound of the horn.

Moving on, you want to avoid using mechanical doors that make noise as they open. If doors squeak, you can use rubber stoppers to minimize the sound.

Finally, you can take advantage of the calm ones among your cattle. Seeing as the calm ones are more likely to turn their ears toward the sudden loud noise than to show agitation, you can look toward their ears to figure out what the source of the noise is. And removing that source should help the other cattle remain calm.

Touch and Cattle Behavior

Probably the most well-known facts about cattle are that they are herd animals. And the implication is that they usually move in groups. As a result, cattle and other animals that travel in herds are accustomed to the feeling of having bodies around them.

Furthermore, cattle are sensitive to touch. And much like they do not like sudden noise or sudden movement, they certainly do not like to be touched unexpectedly. They can react violently if they misinterpret a particular touch as harmful. This is especially true if they have a history of being abused.

In building your chute or any other structure that your cattle must walk through, it's important that you make it narrow enough they can feel the pressure of bodies against them as they walk.

Aside from the fact that the feeling of the nearness of other bodies helps the cows to stay calm, they cannot turn around and go in the opposite direction. So, building a compact chute allows you to kill several birds with one stone.

As for touch, use firm strokes on your cattle. Firm strokes communicate to them you are not only intentional, but that you also mean well. Avoid using uncertain or sudden strokes as those could convey the exact antithesis of what you want them to. Pats are something you want to avoid as your cattle could confuse them for hits.

Finally, when dealing with day-to-day activities, it's important to consider whether any of your cattle have been abused because cattle rarely forget the abuse, and this might translate into their overreaction to your touch.

If any of your cattle have been abused you have to be careful and, more important, extremely patient and calm when touching them.

Health and Cattle Behavior

An animal that isn't at peak health could be difficult to manage. Unfortunately, a lot of factors can make your cattle sick. Many of these health issues are things you have control over, and a few others aren't.

For one, extreme weather conditions can put your cattle under the weather. And for cattle, heat is a bigger threat than cold. Other factors that can affect your cattle you have little control over include parasites, diseases and predators.

However, there will be things you have control over, like food. Your cattle must be properly fed (food and water) if you want them to be agreeable. But you do not want them to be overfed as that can make them bored and lethargic, making it difficult for you to get them to do anything.

Undergoing certain procedures can also affect your cattle's response. For example, having your testicles ripped off will bring any man down and make them uncooperative. So, expect that from your cattle that have just been castrated!

As much as it's within your control, make sure your cattle are in optimal health. And even those factors you can't control, you can try to manage. Here are a few tips:

Feed your cattle promptly and appropriately. Give them enough food as and when due.

Cattle don't take well to being isolated. So house them in herds.

If you just castrated your cattle, you must allow them time alone to heal. They might need to be isolated during this period, but make sure they are in a place where they can see other cattle.

Remember the high wall slab mentioned earlier? The one that should help keep out distractions? It could also help to keep out predators. Having a herding dog can also be useful.

The Flight Zone

From all that has been mentioned, it's easy to conclude that it doesn't take much to spook a herd of cattle. In fact, because of how easy it is, you must pay attention to their reactions.

In simple terms, the flight zone is your cattle's personal space. It is the distance and space from the cattle you can stand and move within without making your cattle becoming violent.

The interesting thing about the flight zone is that getting into it makes the cattle walk away from you. And you could use this to get them to where you want them to go.

To get a cow or bull to move forward, you must stand at the edge of its flight zone. And to get it to stop moving, you must get out of its flight zone but make sure that you are still within sight.

Now, the flight zone distance differs from one animal to another. And this difference could result from temperament or even training.

The calmer an animal is, the shorter the distance of their flight zone, if docile cattle have almost no flight zone (which would make herding them difficult). Also, as your cattle get used to having you around, they get more comfortable and this will diminish their flight zone distance.

Now, how do you figure out your cattle's flight zone? Just try walking gently toward them within their line of sight. When they walk away from you is the edge of their flight zone. If you keep getting closer, your cattle might get really agitated. So, you want to work with that spot where they moved away from you.

Also, if you want them to move forward, use the flight zone behind their forelimbs as they will try to distance themselves from you by moving forward. If you come from their front, they'll most likely go back to where they were coming from.

Finally, cattle can see 300 degrees around them. So, their blind spot is directly behind their head, and that is a place you do not want to be.

Remember that cattle do not like surprises, so standing in a spot where they can't see you but can feel your presence will spook them. And cattle spooked like that can get aggressive, leading to serious injuries for you.

Chapter 4: Facilities, Housing and Fencing

Now, it's time to think about where your cattle will stay and what facilities you'll be using for your operation. And it goes without saying that these housing and operating facilities must be in place a couple of days before your cattle arrives.

However, the facilities you need to operate your cattle rearing outfit depends upon the outfit you want to run. Basically, there are three types of cattle rearing outfits:

- Cow-Calf Outfit: You are breeding cattle.

- Feeder Outfit: You are raising cattle to be sold as meat.

- Combination Outfit: A combination of both.

While a cow-calf rearing outfit requires roofed living quarters for the cattle, it's unnecessary for a feeder outfit. And while a feeder outfit requires a lot of confinement pens and automated feeding systems, a cow-calf outfit doesn't need as many pens. As far as the handling facilities, you'll need the same things no matter what sort of cattle rearing business you want to run.

With all that said, let's see what facilities you'll need to run a successful business.

Beef Cattle Handling Facilities

Headgate

The headgate is a gate device used to hold the head of cattle in place. The idea is to hold the animal in place so the cattle can be accessed to receive veterinary treatment.

There are four broad types of headgates. The self-catching kind automatically closes immediately after the cattle step in, but if that doesn't work too well for you, you can get a version that can also be operated manually.

The scissor stanchion comprises two pieces with a pivot at the bottom. The full opening stanchion comprises two pieces that slide open to allow the cattle in and then slide shut to keep them in. The positive control headgate, which isn't very safe, locks the standing cattle in firmly (but maybe too firmly).

In choosing a headgate, you want to remember the following things:

● An automatic headgate usually locks either too tightly or too loosely. So, it might not always be a good choice.

● If you run a small farm with agreeable cattle or the cattle you want to work on are sick, the automatic headgate will work well.

● However, if you want to get an automatic headgate, it's best to get the kind that allows you to run it manually.

Holding/Squeeze Chute

The holding/squeeze chute is typically attached to the headgate. Many people do without this and just stick with the head gate. If yours is a small outfit, that could work.

But if you have the cash to spare, the holding chute is a good idea. It basically holds the rest of the cow's body so you or the vet can work on the animal without the risk of injury to the cattle or the handler.

Working Chute (Race)

The working chute is a passageway from the crowding pen into the squeeze chute or the headgate (depending on which one you have). A working chute is usually wide enough for just one head of cattle. So, your cattle must pass through the chute in a single file.

Holding Pen

The holding pen is that area where your cattle will stay, pending being taken through the working chute and into the squeeze chute.

Your holding pen should be able to hold as many cows as you'll be working on in one session. This means that if you have a five headgate - holding chute combos (or singles), the pen should accommodate five cows. The reason for this? So your cattle do not get restless waiting in the holding pen.

Loading Chute

The loading chute is used for moving cattle to a trailer. This loading chute should be able to move the cattle quickly enough so those that have entered the trailer do not get restless and moving about.

Also, your loading chute must hold as many cattle as your truck can accommodate.

A loading chute looks like a mobile passageway with raised walls and comes in various sizes.

Scales

You also want to consider getting scales. While you might not need scales, if you are not running a commercial outfit, you'll need them for weighing feed and the calves when they are born.

If you'll be weighing your cattle, placing the scales close to your chute system (basically every facility we just talked about) would be more effective than anywhere else. This way, you can weigh them before you work them.

Beef Cattle Feeding Equipment

Feed Trough

A feed trough is usually a rectangular trough into which you'll pour feed for your cattle. It is typically wide enough to allow several cattle to feed simultaneously.

There are different kinds of feed troughs made of different kinds of materials; plastic, wood, and even metal. If your feed trough is outside, plastic and wood are good choices.

Whatever material you decide on, though, make sure that the trough is tough. Recycled car tires are good material for feed troughs because they are tough and won't harm your cattle.

If you have a lot of cattle of different sizes and ages, it's important to get more than one feed trough.

The idea is to be able to feed your herd by using different troughs according to their age and/or size. This is important because cattle of different ages and sizes have different feeding needs.

By feeding them in different troughs, you'll be able to feed them according to their various needs. Also, the bigger cattle won't be able to oppress the smaller ones.

If you are raising feeder cattle, consider getting automated feeders. Feeder cattle must be fattened up, which means they must eat a lot.

Using automated feeders helps ensure that your cattle are fed often, on schedule, without causing you extra stress.

Feed Carts and Scoops

You need carts to move the feed from where it's kept to the trough. You also need a scoop to put feed into the trough. But you can just carry a bag of feed and pour it directly into the trough if that works better for you.

Hay Feeder

A hay feeder is optional if you already have a feed trough, but it's a good idea. Hay feeders typically look like the bed of a truck. There is a handle attached to the feeder that releases the hay.

That said, other kinds of hay feeders that are smaller and more portable; a few are even collapsible. This is a better option to start a small farm.

Water System

You also need to have a water trough on your farm. Now, this water trough is basically a large water bowl with a few additions, and it has quite a few benefits.

First off, a good trough will have a ballcock valve (the ball-like thing in your toilet tank). The ballcock valve helps to control the water supply. Water will come in, but because of the ballcock value, the water will stop at a set point to prevent waste.

Something else you'll find is a stop valve. The stop valve stops the flow of water into one trough. You'll need this if you want to clean out just one trough and leave the others active.

A check valve is something to look out for if your water line is connected to the main house water line. The check valve will help make sure that water from the trough doesn't find its way into the main waterline of the house.

Here are a few other things to think about in picking out water troughs:

- The water trough should be easy to clean. Troughs rarely have a drain, so you cannot just pull out a plug and easily empty the water. You also must consider that you need to clean your trough as often as every three months. So, pick a trough easy to clean.

- Since you must clean your trough, consider getting more than one, regardless of how many cattle you have. This way, you always have a trough full of fresh water for your cattle, even when one is being cleaned.

- Trough size is an important consideration. When picking the right size, you want to consider how many cows you have.

- If you have only one animal, you want to pick a trough just a little bigger than its head so it will not accidentally fall into the trough and drown.

- If you have more than one cow or you have a herd, consider getting a trough about 20 centimeters in depth.

- The water supply hose should be large enough to supply water quickly into the trough so the trough will not run out of the water as your cattle drink. Cattle can become angry if the trough is out of water.

- The water pipes must be covered so your cattle do not trip and fall and/or damage the system.

- Also, make sure that the trough is anchored firmly to the ground.

Beef Cattle Housing

Cattle housing doesn't have to be complicated. If you have a grassy area or a pasture and your cattle are feeder cattle, they can just live in the field in the summer. But it doesn't hurt to have something constructed for them.

Now, here are a few things to remember in constructing housing for your beef cattle:

1. One of the most important things is to construct something that is easy to clean. To do this, you can either make sure that your barn or cattle house has a drain, or you can erect the structure on an elevated area.

Either of these will make it easier to drain the floor while cleaning or if it rains into the structure.

If you'll be installing a drain, make sure that it isn't in an area where your cattle will be walking. Remember, they are not good at depth perception.

2. Make sure that space is properly ventilated and well-lit with natural light. If the entrance to the shelter faces south, that will work well.

Avoid artificial lights as much as possible as cows do not like areas too brightly lit. Allowing for natural lighting will ensure that the cattle are receiving this important light in a uniform manner.

3. Make sure that the ground is level and uniform because various or changing textures can be stressful for cattle.

4. Create a separate space for birthing.

There are several housing options you can consider.

The first choice is an open-sided plan or the closed-sided barn structure. If you chose the open-sided structure, erect a small fence around the sides. It should be small enough so it doesn't obstruct the view, but high enough to prevent your cattle or other outside animals getting over it.

Whichever style you use, the interior is important. Inside the cattle house, it's important that each head of cattle has its own space and doesn't roam freely.

If you have a small amount of cattle and enough space, one row of stalls should be fine. Make sure that the cattle have enough space to be in their stalls. Consider a 105-centimeter side space and 165-centimeter standing space.

But if you do not have that much space or you are running a large outfit, consider doing two rows of stalls, with the cattle facing each other.

Now, if you want to run a fixed housing system where the cattle remain indoors all day, each stall must have its own feeding and water trough.

If you are running a two-row system, on the other hand, the two cattle facing each other can share a feeding and water trough.

Fencing

If you'll be running an open housing system, it is very important that you pay attention to fencing. In the open housing system, your cattle spend most of their time outside, feeding, grazing and resting.

You want to make sure that they do not wander off or get attacked by other animals. So, here's what you need to know to keep them safe.

Choosing The Right Fence

The type of fence you'll want to use depends on the breed of cattle you're rearing and whether you have a predator problem. Different kinds of fences range from barbed wire with woven wire to high tensile and electric fences.

Most cattle eventually hurt themselves on barbed wire. If you are raising beef cattle, predators might not be a problem because of how large and heavy the cows are. So, you could investigate the high tensile or woven wire fence types.

Fence Post Spacing

The general rule of thumb says to space your fence posts 8 to 12 feet apart. The idea is to have enough fence posts so your fence is solid and braced for impact. The posts help anchor the wire and keep it grounded. Too few posts will give you a weak boundary.

However, if you are using galvanized steel and high tensile wires, you can get away with spacing your posts further than 12 feet apart.

You need corner posts to help solidify your fence even further, which is why you need to bury them deep.

The general rule of thumb is to bury the post to a depth anywhere between 35 and 50% of the height of your corner post.

So, if you have a 6-foot corner post, you'll be digging a hole between 2 to 3 feet deep. Also, the hole must be three times as wide as the post.

Land Ownership

Before you erect your fence, find out where your property starts and stops. This way, you aren't throwing away a part of your land or stealing someone else's.

Consider employing the services of a land and quantity surveyor because court battles because of land boundaries can be complex and complicated.

Chapter 5: Beef Cattle Nutrition and Feeding

Before we get into the conversation about what types of food your cattle should eat, it's important that you understand how their digestive systems work. This will help put things in perspective when we talk about feeding and what kinds of feeds are best for your cattle.

- Ruminant animals such as bovines are efficient at digesting high-roughage feedstuff because of the way their digestive system works. This system consists of:

- Mouth.

- Tongue.

- Salivary glands.

- Esophagus.

- Stomach (made up of four compartments viz: the rumen, reticulum, omasum, and abomasum).

- Pancreas.

- Gallbladder.

- Small intestine (made up of the duodenum, jejunum, and ileum).

- Large intestine: (made up of the cecum, colon, and rectum).

On average, cattle will take between 25,000 to 40,000 prehensile bites each day as they graze. Sometimes, it could even be more. As you can probably deduce, they chew fast and don't chew sufficiently before swallowing.

Grazing for cattle takes up more than a third of their day, while the remaining two-thirds are shared between chewing cud (bringing the partially digested food back up to ruminate) and simply being on the land. Chewing cud takes up about a third of the remaining two-thirds, and the rest of the day is spent resting.

Now, here's what the journey of food from the mouth to the anus in cattle looks like.

When the bovine takes in the forage, it mixes with their saliva, which contains potassium, sodium, bicarbonate, urea, and phosphate. This forms a bolus that travels through the esophagus to the reticulum. The esophagus in cattle works bi-directionally, so it moves food downward, but it also pushes the cud back into the mouth. Once the cud is moved back up into the mouth it is chewed and mixed with saliva a second time before it is swallowed again and moves to the reticulum.

From the reticulum, the solid part of the cud goes to the rumen to ferment while the liquid part goes into the reticulorumen. The solid part remains in the rumen for 48 hours. Now, the reticulorumen (reticulum + rumen) contains microorganisms such as protozoa, fungi and bacteria. These feed on the cud in the rumen and break them down into volatile fatty acids (VFAs). Examples of VFAs include acetate for synthesizing fat, propionate for synthesizing glucose, and butyrate. Cattle utilize these VFAs to produce energy.

Now, let's look at each part of the ruminant digestive system in closer detail.

The Ruminant Digestive System

Reticulum

Because of its looks, the reticulum is also called a "honeycomb". Its main role is to move smaller digested food into the omasum and the larger particles into the rumen where these particles are further digested.

The reticulum is also the part of the stomach where heavy objects can get trapped. So, if your bovine mistakenly ingests a wire, nail or any other heavy object, it will most likely get trapped in the reticulum.

As normal contractions occur, the object can pierce through the intestinal wall and move into the heart. This can cause hardware disease, which is why the reticulum is also sometimes called a hardware stomach.

Rumen

Also known as the "paunch," the rumen comes with papillae, the primary tissues through which absorption takes place. It's mostly a fermentation vat because all the microbial fermentation takes place here. It is an anaerobic environment. There is no oxygen there. It has a pH range of 6.5 to 6.8.

The rumen is also the place where gas is produced, which makes sense since it's where all the fermentation takes place. Gases like methane, hydrogen sulfide, and carbon dioxide are produced in the rumen.

Omasum

The omasum links to the reticulum via a short tunnel. It is spherical and characteristically comes with many flaps or leaves, which is why it is also called the butcher's bible or "many piles." The omasum is where water absorption takes place for ruminants. In cattle, the omasum is large and well developed.

Abomasum

Of the four compartments, the abomasum is the actual stomach because it is the most like non-ruminant stomachs. It is highly acidic, with a pH range of 3.5 to 4.0 but the animal is safe because the cells of the abomasum secrete mucus, which protect the abomasum from acid damage.

The abomasum also contains hydrochloric and digestive enzymes such as pepsin and pancreatic lipase, which work together to break down food.

Small and Large Intestine

In the small and large intestine, nutrients are further absorbed. The small intestine is long (about 150 feet), has a 20-gallon capacity, and is even more acidic than the abomasum. From the abomasum, digested food moves into the small intestine. When this happens, the small intestine becomes alkaline as pH increases from about 2.5 to 7 or 8. The increase in pH is necessary for enzymes of the small intestine to act properly.

Just like in the small intestines of humans, there are villi in the small intestine of cattle. These villi look like fingers and increase the surface area of the intestine to help the absorption of nutrients. As the muscles contract, food is moved from the small intestine to the large intestine.

The main function of the large intestine is to reabsorb water from the food digested while it passes the rest into the rectum.

Nutritional Requirements of Cattle

There are several nutrient classes cattle need for their bodies to develop and operate properly. Each nutrient class has its own role to play in the body, and an absence or deficiency of them could inhibit growth or cause ill health.

The following are nutrient classes that cattle require:

TDN (Energy)

It's obvious what energy does in the body of any living organism. This energy gives us the drive to carry out work and, for living organisms, work includes growing, lactating, reproducing, moving and digesting food.

The energy in cattle nutrition is expressed as Total Digestible Nutrients (TDN) and is the most important nutrient that cattle need. They need it in large amounts too.

For cattle, energy sources include hemicellulose and cellulose from grain starch and roughage. They can also get energy from fats and oils, but these only make up a small part of their regular diet.

Protein

We know proteins to be the building blocks of the body. They form the main components of organs and tissues in the body, such as muscles, connective tissue and the nervous system.

A protein comprises several units of amino acids linked to form chains. When supplied in adequate amounts to the body, it helps normal body maintenance and in lactation, growth and reproduction.

The different components of protein vary in their solubility. There are the digestible proteins digested by microbes in the rumen, and then there are the insoluble proteins that leave the rumen intact to the lower gut.

Minerals

There are macro-minerals and micro-minerals. Macro-minerals are needed in a relatively larger quantity than micro-nutrients. Examples of macro-minerals include calcium, sodium, phosphorus, potassium and magnesium. Micro-minerals, on the other hand, are also known as trace minerals, include copper, iodine, selenium, zinc and sulfur.

How rich in minerals your cattle's diet is depends on the quality of feed they are consuming. Often, you must fortify their ration with mineral supplements. The type of mineral supplements you choose also depend on the feed your animals are eating, and their nutritional requirements.

Minerals are a critical part of your cattle nutrition, and even though they are only needed in relatively small quantities when compared to other nutrient classes, a deficiency can have mild to moderate to severe consequences. A few of these consequences include poor growth, bowed legs, brittle bones, a fall in conception rates, muscle tremors, convulsions, etc.

Vitamins

Vitamins are like minerals in their function. For beef cattle, the most important vitamins include vitamins A, D, and E. Fresh foliage is a good source of these vitamins. While older forage contain vitamins, vitamin levels do tend to drop after a while. Silage and grains also have lower levels of vitamins.

Vitamin A ensures normal reproduction, growth and body maintenance. Vitamin D is necessary for the proper development of bones. With selenium, vitamin E ensures that muscle tissue develops properly.

Absent these vitamins, cattle can experience reduced fertility (vitamin A deficiency), rickets (vitamin D deficiency), and muscular dystrophy, and white muscle disease (vitamin E deficiency).

White muscle disease is a common problem with cattle. To prevent this, you may have to inject the calves with selenium or vitamin E at birth. Feeding your cows supplementary selenium/vitamin E or injecting the pregnant cows with selenium/vitamin E can also help.

Vitamin B has little impact on cattle nutrition. The microorganisms found in the rumen already produce this vitamin in sufficient quantities, which the cattle absorb. But vitamin B is essential for calves, as they haven't fully developed their rumen yet. Super stressed cattle might also need vitamin B supplements as stress depletes the microbial population in the rumen and thus diminish the vitamin B.

Now that you know your cattle's nutritional requirements, let's go into the different categories of foodstuffs you should feed your cattle to nourish them with all the nutrients they need.

Types of Cattle Feeds

Grain Supplement

Grain is rich in energy and has moderate amounts of protein, but it contains little fiber.

But grain is great for cattle because it facilitates rapid growth and helps to fatten your cattle. Providing grain is a feeding method adopted by most farmers because of its cost-effectiveness.

Grain is also a fantastic alternative for cattle-rearers who live in areas where access to excellent hay is limited. In the winter, too, grain can be a lifesaver for farmers and cattle.

Although grain has excellent benefits, it's important to not let your cattle get too dependent on it. Cattle dependent on supplements reject pasture and hay, which are much better options for them nutritionally compared to supplements.

Examples of grains include barley, corn, and oats.

Roughage

Examples of roughages include hay, grain hulls, grass and oilseed hulls. Roughage is typically rich in cellulose and hemicellulose (fiber) but pack little energy. It supplies moderate levels of energy. It does contain protein though, depending on the plant from which it is derived and the plant's level of maturity.

Now, since we're here, let's talk a little about hay.

Hay is one of the best feeds cattle can eat. It can single-handedly supply almost every nutrient cattle need, but it must be eaten at the right time or you lose all the dense nutrients. In other words, you should pick it before it dries. Also, proper curing and storage are very important when it comes to feeding your cattle hay.

Hay comes in different varieties, each with the level of nutrition they offer. Alfalfa, for instance, is richer in phosphorus and calcium than grass, but grass hay has high levels of protein. Hence, most experts recommend that you mix alfalfa hay with a bit of grass as against feeding alfalfa exclusively, especially when raising beef cattle.

Alfalfa is great and is even recommended for dairy cattle, but because of its tendency to cause bloating, it is not recommended for beef cattle. So, for your beef cattle, you can mix alfalfa with grass hay, or feed them legume hay, which is protein-rich.

Forage and Pasture

Pasture and forage crops contain all the nutrients cattle need to thrive. Unless the soil is depleted for one reason or another, or it may be too early in the year for grass to grow lush and rich.

Besides grain, forage crops and pasture are other inexpensive feeding solutions for your cattle, but you must do your due diligence before feeding them just pasture and forage crops. Knowing the fertility of the soil and ensuring good watering is important to ensure the plants are packed with adequate nutrients.

Also, always know what kind of plants your cattle are eating, and their condition and maturity level.

Oilseeds

Oilseeds are rich in protein and energy, but their fiber content varies. Examples of oilseeds include canola meal and soybeans.

Byproducts

By-products come with high moisture levels, and their nutritional content varies depending on their source. Examples of byproducts include sweet corn cannery waste, distiller's grains, grain screenings, apple pomace and bakery waste.

Chapter 6: You Can Still Milk Your Beef Cows!

The cows that bless us with ice cream and the cattle that provide us with delicious steak are different creatures. But people eat *dairy cow meat* and drink *beef cow milk*! Before we venture into *that* conversation, it's important to lay the foundation for this discussion. Hence, we will kick off this chapter by looking into beef cattle and dairy cattle to understand the similarities and differences between these two breeds. We know you're here for the beef, but knowing something about milking and dairy cattle is also important.

Beef vs. Dairy Cattle

Beef Cattle

Beef and dairy cattle look characteristically different. Beef cattle look stocky like bodybuilders. They channel all their energy into storing fat and developing muscles. These work together to give you delicious beef; the best is lean meat with marbling for enhanced flavor and texture.

Beef cattle's strong legs help them navigate pastures. Their bellies are also rounded and stocky with thick backs, strong shoulders and rumps, and short necks.

With diet, beef cattle feed primarily on grains and grass, although they eat more grass than grains, especially when they are still young.

You probably already know that beef cattle produce milk because how else would they nurse their calves, right? But as you can also probably deduce, milk production in beef cattle is much lower than in dairy cattle. The logic is simple. Over the years, beef cattle have been bred to do one thing, and that is to produce beef. So, though beef cattle produce milk, they only produce enough to nourish their calves, which only yields about a gallon or two daily.

Beef can come from a steer, a cow, or a heifer, but the best beef comes from heifers and steers.

Dairy Cows

If beef cattle are like bodybuilders, then dairy cows are like marathon runners. They might look underfed, but that's how dairy cows are genetically wired. No matter how well-fed they are, they remain lean and angular because they channel all their energy into lactating rather than building muscle or storing fat. For cows, milk production and bulking up in mass are mutually exclusive. Therefore, dairy cows and beef cattle are characteristically different in appearance.

Dairy cows produce milk in large quantities daily, up to ten gallons per day usually. To keep her healthy and comfortable, you must milk your dairy cow two or three times a day.

Dairy cows are raised in pastures or free-stall barns where they get access to fresh water and food. They have the same diet as beef cattle, which consists of grass and grains, but unlike beef cattle, they need not navigate the terrain to graze, hence their slight build.

Now, remember that cows will only lactate when they have calves, and cows can have only one calf in a year — beef or dairy. Milking then occurs for roughly 300 days in a year, after which the body takes a break for the remaining 60 plus days as they prepare to calve.

However, let's move into the second part of this chapter. Let's talk about milk production in beef cattle.

Milk Production in Beef Cattle

While beef cattle are primarily raised for their meat, nothing says you can't milk them when they lactate. The taste differs slightly from what you get from dairy cows, and the quantity is also not as high. Nonetheless, even if it's not for sale, you can still milk your beef cow and enjoy the dairy with your family.

That said, milking beef cows have their benefits and their disadvantages.

Why High Milk Production in Beef Cows Is Great

If a beef cow produces plenty of milk, her calf will be sufficiently fed and nourished. This is great for the calf's health. It's also great because calves that are fed sufficient milk early in life attain a heavier weight by weaning time. A study by the state of Oklahoma confirmed that more milk translates into an extra 30 pounds of weaning weight for calves.

Still, High Milk Production in Beef Cattle Has Its Disadvantages

Even with the benefit of a heavier weaning weight for the calves, there are still important reasons farmers might prefer lower levels of milk production in their beef cows.

During times of nutrient deficiency, a cow's body will channel the energy generated into three main areas: body maintenance, lactation, and reproduction. Now, look at these areas as levels of a sort. In other words, if the demands of one level are not met, energy will not be supplied for the next level. Hence, body maintenance is a priority, and only if the energy requirements are met will energy be supplied for lactation. Then, only when energy levels for lactation are met will the cow be biologically prepared to breed.

So, it's easy to deduce from all we're saying that milking-cows need a lot of energy. You'd need to help them keep up by giving them feed in large quantities for them to keep producing milk in sufficient quantities. It's nearly impossible to raise high milk-producing cows in a grass-based, low-input system. Heavy milkers have a poorer body condition when compared with their counterparts producing moderate amounts of milk.

How to Choose the Best Beef Cows for Milk Production

Picking a beef cow based on her milk-producing ability requires careful thought and consideration. If you go for a heavy milker, you'll get a heavier calf, which is great because they are more valuable. But a heavy milker is way more expensive to maintain than a moderate milker. You must be prepared to provide supplemental feeding, and that will increase your expenses.

Below is an estimate of how much food various cows in the early stages of lactation need:

- 10 pounds of milk daily: approx. 26.5 pounds of dry matter daily.

- 20 pounds of milk daily: approx. 29.0 pounds of dry matter daily.

- 30 pounds of milk daily: approx. 31.5 pounds of dry matter daily.

So, if you're sure you can get cheap surplus feed, then why not invest in a heavy-milking beef cow? But if feed is expensive, it might be better to stick to one that's just moderate in its milk production.

Dual-Purpose Cattle

As the name probably already suggests, dual-purpose cattle are bred both for their beef and for their milk. In times past, cows were triple-purposed — milk, beef, and draft work. But then horses came on the scene, and cows could take a break.

Dual-purpose cows are not as popular as they once were because farmers preferred more specialized breeds, especially because it confused the breeding purposes. Questions arise such as "What exactly do we want from this cow, and how do we raise her?" Owing to this confusion, farmers manipulated the gene pool to produce beef cattle and dairy cattle specifically through breeding.

Today, there are still a few dual-purpose breeds to be found in the cattle-raising community. But they are limited to small farmers and small farmlands. Dual-purpose cows are not to be found in the specialist dairy industry. They are better suited for the small farmer because they produce more protein, more fat, and more liters of milk, all at lean body weight.

Dual-purpose cows work well for small farms because they make ideal house cows. They will provide enough milk to feed your family comfortably and still make enough to feed their calf. They also hardly ever grow to the standard, gargantuan cattle size. Hence, they are easier to house, and require a less land than regular beef cattle.

You can crossbreed cattle to create a dual-purpose cow. If you do, though, it's important to keep it to the first cross or the second one at the most. Crossbreeding produces hybrid vigor, which gives the offspring the genetic advantage it has over both its parents, but it waters down the more you crossbreed; hence, the reason you should never go beyond the second cross.

How to Milk a Cow Correctly

Use Clean Equipment

It's best to use a stainless steel bucket to collect the milk as it's easier to clean and disinfect. Also, always make sure that all your tools and equipment are always 100% clean.

Tie Up Your Cow

If you don't tie your cow up, she will easily walk off while you milk her in search of the pasture where she can graze or wherever else catches her fancy. A good way to tie your cow in place is to use a neck collar or a halter.

Now, make sure your girl has a snack waiting to be given to her after being a good girl during milking. You can give her hay or a small quantity of grain as a snack so she will cooperate with you and enjoy the experience of being milked.

Also, ensure that her stall is a place where she feels comfortable, so she enjoys going there. It would make your life much easier.

Prepare the Udder

Before milking, wipe the udder with a warm rag to remove dirt, manure, hair, and debris. This is important to keep the skin on the udder from drying out and cracking. If the udder looks dried out, you can apply a moisturizing dip to rejuvenate the skin, making it easier to milk her when the time comes.

Strip Each Teat

To confirm that the milk is okay, you can squirt the first few drops of milk into a cup or onto the ground. It's called stripping the teat. Milk should be smooth and white without clumps when expressed.

Now, you want to do this stripping for each teat. Once you've confirmed that the milk is good, you can go to the next step before milking into the stainless steel bucket.

Apply Pre-Milking Disinfectant

After stripping the teats, apply pre-milking disinfectant to the teats, following the manufacturer's instruction. Once you finish applying it, clean off the disinfectant with a clean, dry towel.

Express the Milk

Milk from the teats in the front quarters and squeeze them alternatively until they are both empty.

To milk your cow, hold up your hands to the teats in the front quarter first, like you're holding a cup to drink out of. Then hold each teat between your forefinger and thumb and squeeze them to get the milk out. Keep doing this until the udder is empty. You can easily tell the udder is empty because it will become flaccid.

Once you've ensured that you've milked all the teats on that udder, apply post-milking disinfectant. Voila! You're done!

Chapter 7: Beef Cattle Hygiene, Health, and Maintenance

One very important reason to pay attention to hygiene in your cattle rearing practice is that it has a huge impact on the health of your cattle. In addition, hygienic practices are important for your customer's health.

However, we do acknowledge that many bovine diseases might not result directly from poor hygiene. But whatever the cause, poor hygienic practices can exacerbate the situation.

Whichever way you look at it, it is very important to keep your cattle and their living quarters clean. And as for ensuring that they're healthy, you wouldn't have a practice if your cattle were sickly or dead, anyway.

Beef Cattle Hygiene

In looking at how to maintain hygiene in your beef cattle practice, we'll be breaking things down into the three commercial life stages of cattle: rearing, housing, and transportation.

Rearing

There are several practices involved in the rearing of cattle, or what many call cattle husbandry. You'll have to feed them and clean them.

You might also need to perform (or have a professional perform) medical procedures like embryo transfer, artificial insemination, birthing, or castration. All these must be done hygienically. Let's see how.

Hygienic Cattle Feeding

• The Agriculture and Rural Development Department in Namibia recommends that you wash your cattle's water trough once in three (3) days. As for your food trough, you can wash that out once a week.

• Before you wash your troughs, you must empty them first. You'll need a scoop to get the water out because food and water troughs usually do not have drains. When you're finished getting out the content, spray the trough down, add in dish soap, and start scrubbing.

• If you can afford it, consider getting new water and feed troughs instead of used ones. New troughs stay cleaner longer and are slower to form algae.

• Keep the water in the trough fresh. Algae and amoeba are not things your cattle should be consuming. So, once you notice the water changing colors, it is time to throw it out.

• Keep your water and feed troughs far from each other. If the feed trough is close to the water trough, your cattle could get feed into the water and water into the feed, causing both to become useless quickly.

Cattle Cleaning

• You can set up a sprinkler system that will spray your cattle with water regularly. You can also set this system up as part of your cattle caretaking system so that all your cattle are sure to pass by regularly.

• But having your cattle sprinkled with water, no matter how regularly, will not be enough. So, you must scrub them once a week.

• Before you scrub any of your cattle, make sure that it is properly held down. For this, you'll need a separate washing head chute so the animal's head is secure while you scrub the body. Be sure that you are not using your regular head chute so it is not perpetually wet.

• Purchase special cattle brushes and shampoos. Do not use all-purpose scrubbing brushes or human shampoo or you can harm them.

• When you wash your cattle, work your way from top to bottom.

• You should also rinse them from top to bottom. And as you rinse, slide your hands down their body to be sure that all the soap suds are gone.

• Cattle prefer to be bathed on a warm day, so consider bathing them when it is warm outside. Not only will they be more comfortable, but they'll also dry faster.

Hygienic Medical Procedures

If you employ the services of a veterinarian for all your medical procedures, the medical procedures will always be hygienic.

But it doesn't hurt to know what hygienic procedures should look like. Plus, we acknowledge that running a small practice might require you to do medical procedures on your own. So, here are several things to note.

- Make sure that every cow is vaccinated yearly. If you try to breed a sick animal, you're setting your practice up for failure.

- Make sure you always have a first aid kit handy. And sterilize your kit and other items after every use.

- If any of your cattle falls into a ditch and is severely injured or injures itself in another way, immediately tend to the wound. You do not want to risk an infection.

- Create a separate space for birthing. It need not be in a different room if you do not have the luxury of space, but it must be away from the regular living quarters. Also, furnish the space with clean, fresh straw before your cow gives birth.

Housing

As much as cleaning your own living quarters is important, cleaning your cattle's living quarters is also important. It is probably more important, considering that you don't poop everywhere, but your cattle do!

If you have live-in pets, you might have a little idea of what it takes to clean out your cattle's live-in quarters. With that in mind, let's see what needs to be done:

- One of the most important things you must do in cleaning out your cattle's living quarters is to shovel out the poop. Your cattle should not be living in their own filth. So, it's normal practice to shovel out the poop every day.

- Shoveling out the poop will not be enough, though. You'll have to go a step further to disinfect the floors. Make sure that whatever disinfectant you use is safe for both you and your cattle. You also want to be sure that the disinfectant is versatile (can kill a variety of bacteria) and works quickly. It should also not contain components that are not compatible with certain building materials.

- If you use straw bedding, change it out once every four or five days. If you use sand beddings, dig it out and change the sand when you notice a dark layer of sand across the top.

- If your cattle eat in their stalls, you might need to clear out the stalls every day to rid them of the droppings.

- Make sure that your cattle have ample living space, as squeezing them in will lead to the quick spread of infections, diseases, injuries, and respiratory problems and is generally just uncomfortable.

Transportation

- You need to make sure that the trucks you use to transport your cattle are regularly cleaned out. You need to clean out your truck after every trip.

- Do not transport sick cows with healthy ones. If you need to take one to the vet because they are sick and another to be vaccinated, you might need more than one trip for that.

- Make certain that your trucks are properly ventilated.

- Make sure that your cattle are not left in the truck for long periods of time and that they have access to food, water, and fresh air. Portable feeders and water bowls come in handy.

- If you are transporting newborns, use a disinfected wheelbarrow or calf taxi. It is best to get a new calf taxi to transport your new calf.

Beef Cattle Health

While proper hygiene will help keep your cattle healthy and well, vaccinations and an understanding of the health issues cattle are prone to are also important. Both topics are what we'll be exploring in this section.

Cattle Health Issues

Your cattle could fall prey to diseases, parasites, and food poisoning. Let's look at these specifically:

Common Cattle Diseases

Tail Rot

Tail rot is what it sounds like: rotting of the tail. It is most probably a result of an animal continuing to use its tail to swat at flies, although its tail has been injured, broken, or dislocated.

Tail rot is more prevalent in wet areas, and during the rainy season as floors become slippery. It is also more prevalent in places with a lot of trees because cattle can hit their tails violently against trees as they walk past.

To Prevent Tail Rot: You must get rid of everything, and every situation that your cattle can hit their tail against or that could trip them. Also, vaccinating your cattle against tetanus will make sure they are not susceptible to tetanus if they do break their tail.

To Treat Tail Rot: You must get the animal vaccinated against tetanus. Also, the animal might need to get its tail amputated. If blood flow to the injured part of the tail is completely blocked, amputation might not be necessary as it will eventually dry up and fall off.

Akabane

Akabane is a disease that causes deformities in cattle fetuses. It is caused by an arbovirus, and it has no clinical symptoms. It is spread by blood-feeding insects (most commonly midges), and it affects the nervous system of a fetus.

To Prevent Akabane: The only way to prevent akabane is to kill the midges in your area. Also, exposing a herd to a place where akabane is endemic can help the herd gain immunity.

Botulism

Botulism is a bacterial disease that affects cattle and is caused by the Clostridium Botulinum bacteria.

The bacteria thrive in decaying plants and animal carcasses and in moist environments. It produces spores that, if in the right environment, will survive for a long time.

An animal can get infected by consuming anything that has been infected with the spores or has come in contact with infected carcasses.

Symptoms include paralysis of the facial muscles and the limbs. And death can occur 1 to 14 days after the first symptoms.

Humans can get a botulism infection, too, but this will not result from contact with an infected animal but from consuming infected food and/or drink.

To Prevent Botulism: Make sure your cattle are promptly immunized against botulism. Also, properly dispose of carcasses and bones from your property.

To Treat Botulism: If you immediately notice that an animal has consumed an infected substance, purging that animal might work, but the prognosis for botulism isn't good, and infected cattle usually die.

Stringhalt in Cattle

Stringhalt in cattle is a knee dislocation where the inside ligament hooks over the knee at the top. The affected leg will be straight, and the animal must drag that leg until the ligament releases, and the animal can walk freely.

Stringhalt is almost always genetic, where the animal has an anatomical defect in the leg. And while poor nutrition could make the condition more evident, it is usually not the cause.

If it happens suddenly, stringhalt could result from injury or a phosphorus and calcium deficiency.

To Prevent Stringhalt: Do not breed with bulls that have stringhalt. Now, it might be difficult to detect this in cattle that are of optimal health, so you must make a very careful selection when time to breed.

To Treat Stringhalt: Many cattle-rearers take their cattle in for surgery to treat the affected knee, but most people put the animal down.

Three-Day Sickness/Ephemeral Fever

The three-day sickness is a viral disease transmitted by mosquitoes. It is prevalent during the wet season when mosquitoes have ample opportunity to breed.

The three-day sickness usually presents as mild signs like a fever, temporary lameness, and eye and nose discharge, moderately severe signs like swollen joints, depression, and subcutaneous swelling, and severe signs like paralysis and a coma.

Most times, these symptoms disappear after three days, and the affected animal is back to normal, but there is also a significant possibility that the affected animal will die before the symptoms go away.

To Prevent Ephemeral Fever: a vaccine can be administered to your cattle to keep them immune. They'll have to take two doses, four weeks apart.

To Treat Ephemeral Fever: Seeing as animals usually recover on their own, the best you can do is to ensure that they are comfortable, properly hydrated, and well-fed.

Common Cattle Parasites

Ticks

Tick fever is caused by exposure to these blood parasites. Tick fever can be deadly, and if it isn't, it can lead to other complications like abortion of a pregnant cow, infertility for a period with bulls, and eventual financial loss for you.

Cattle that have tick fever might experience a loss of appetite, general body weakness, and/or depression. Cattle between the ages of 18 and 36 months are more prone to a tick fever infection.

To Prevent Tick Fever: If cattle are exposed to the parasites between the ages of 3 and 9 months old, they might develop a long lasting immunity against tick fever.

To Treat Tick Fever: If you suspect that any of your cattle have tick fever, consult your veterinarian for diagnosis and treatment.

Worms

If worms have become an issue with your herd, it might be difficult to recognize because the outward expression of symptoms is similar to poor nutrition.

You want to gauge the egg per gram (EPG) of dung. So, if you check your cattle's dung, and there is over 200 EPG, you might have a problem on your hands.

If so, you might have to resort to drenching, which is administering certain chemicals to your cattle to rid their systems of parasites, including worms.

But, seeing as this is a sensitive issue, consider consulting with your veterinarian if you suspect there is a worm infestation among your herd.

Food Poisoning

Grain Poisoning

This happens when cattle consume large amounts of grain they shouldn't have eaten for various reasons. It is most likely to occur if you switch your cattle from pasture to grain or if your cattle accidentally gain access to grain.

An animal with a case of grain poisoning may show a few of these symptoms:

- Loss of appetite
- Depression
- Diarrhea
- Smelly feces
- Increased heart rate
- Bloating
- Eventual death

To Prevent Grain Poisoning: Slowly introduce grain into your cattle's diet. Start by mixing it in small amounts with what you already feed them.

Then progressively reduce the amount of their old food and reduce the amount of the grains before you completely phase out the old. Also, keep the grain out of the reach of your cattle.

To Treat Grain Poisoning: If any of your cattle just ate a large amount of grain and you think there is a risk of poisoning, immediately feeding it hay can potentially help it recover.

Otherwise, you can consider slaughtering, as killing the animal before acidosis develops might be the more financially wise decision.

Urea Poisoning

Urea poisoning is caused by excess and/or irregular consumption of urea. An animal with a case of urea poisoning will show a few of these symptoms:

- Facial muscles are twitching
- Teeth grinding
- Abdominal pain
- Bloating
- Weakness
- Rapid breathing
- Spasms
- And eventual death (usually near the source of the urea)

To Prevent Urea Poisoning: Make sure your cattle do not have access to urea.

To Treat Urea Poisoning: You might have to resort to drenching, but this rarely works. Do consult your veterinarian if you suspect that any of your cattle have a case of urea poisoning.

Cyanide and Nitrate Poisoning

Your cattle can get cyanide and nitrate poisoning from sorghum crops. These crops are generally safe to consume, but they often release toxins in hot weather when they have been stressed.

An animal with a case of cyanide or nitrate poisoning will exhibit a few of these signs:

- Labored breathing
- Bright red mucous membranes
- Muscle weakness
- Convulsing

• Death

To Prevent Cyanide and Nitrate Poisoning: If you suspect that any of your cattle is sick with or has died of cyanide or nitrate poisoning, remove any source of cyanide and/or nitrate in your cattle's feed and consult with your veterinarian.

To Treat Cyanide and Nitrate Poisoning: Consult your vet!

Vaccination

To keep your herd healthy and prevent them from being susceptible to diseases, vaccinate your cattle.

Now, there are several vaccines that cattle need to take for various reasons, and you should consult with your vet to know the ones specific to your herd and location. Generally, your cattle should receive these vaccines:

- **Clostridial Diseases Like Tetanus:** Two shots should be administered 4 to 6 weeks apart as a 5-1 'package.' And you want to make sure your herd gets their first shot as early as 6 months of age. Then you can administer it based on your discretion.

- **Three-Day Sickness:** Two shots should be administered 4 to 6 weeks apart. It is usually too expensive to vaccinate your entire herd, so consider only vaccinating those valuable ones (that is, the ones you want to breed, especially considering you need them to live long enough). Also, you must continue administering the shots every year. Spring is the best time to do so.

- **Botulism:** Depending on the vaccine, you might need to administer one or two shots 4 to 6 months apart. These will need to be administered every year but do not administer this simultaneously you are administering another vaccine.

- **Tick Fever:** A one-time shot. If you are introducing cattle coming from an area where the tick isn't prevalent, administer a second shot to the new cattle. That said, consider administering the shot early, say around 3 to 9 months of age.

Ensuring your cattle's hygiene and health is important for your business, but more important, for the comfort and wellbeing of your cattle. So, pay attention to the things mentioned in this chapter and make sure you have the contact of a trusted vet.

In the next chapter, we go into the specifics of the different genders in your herd.

Chapter 8: Bulls and Steers

While bulls and steers are both male bovine cattle, they aren't the same. And the difference between them was hinted at in the first chapter. But in this chapter, we'll be expounding on those differences and explaining how they affect your practice.

Bulls

Basically, bulls are mature male bovine cattle used for breeding. All male bovine cattle are born as bull-calves.

You will then need to carefully examine those calves to decide if they have characteristics you want to see in your cattle.

If they do, keep them intact and use them for breeding your cows. But if they don't, castrate them.

Steers

Steers are castrated male bovine animals; their testes have been removed while their penis remains intact.

Castrating a male bovine animal that you don't want to breed will keep them from being aggressive, especially when cows are in heat. So, unless it is necessary for you to have bulls, castrate your male cattle.

Now that we've got that settled, let's explore the differences between bulls and steers.

Bulls Vs. Steers

Physical Differences

Bulls are typically the biggest of the cattle, and this has something to do with the amount of testosterone they produce.

Because steers are castrated (and so cannot produce so much testosterone) at an early age, they do not grow to be as big as bulls.

In fact, if not for the fact that steers have a penis while heifers have a vulva, it would be difficult to distinguish between them.

Bulls have a more pronounced penis as opposed to steers and are also bushier around the sheath that covers their penis.

Behavioral Differences

Bulls are generally more difficult to control and keep in check than steers. And things could get even worse if there is more than one bull in the same space. They'll fight one another for dominance and can transfer their aggression to the handler.

Things can escalate even more if there is a cow in the vicinity that is in heat and ready to be bred. Bulls can easily hurt a person that tries to keep them from their precious female.

Now, this does not mean that bulls are impossible to work with. After all, there are practices where bulls are bred. But if you want to raise bulls, special care will need to be taken.

Steers, on the other hand, are generally tamer and easier to handle, especially considering that they are not as big as bulls.

Also, steers have their sexual urges repressed because of the castration. They are less likely to cause fights with other animals or even their handler.

Handling

Because of the behavioral differences between bulls and steers, the way you handle them is different. You'll need help with anything that requires you to touch the bull.

Your bull will walk through the holding pen just fine, provided you are in its flight zone, but getting it t to enter the squeeze chute won't be so easy. You might need someone else on standby in case you need assistance.

It's important to note that with bulls a head gate might not be enough to hold them still. You must – almost always – couple the head gate with a squeeze chute whenever you are doing veterinary procedures on them.

Now, for steers, it is important to remember that while they have had their testes removed, they are not completely immune from aggressive behavior. So, you do not want to treat a steer as if it is harmless.

However, handling a steer is something you can almost always do on your own unless that steer has a terrible temper.

You might also get away with just a head gate, but if you can afford it, consider buying a squeeze chute – even if you are handling a steer.

Quality of Meat

Because steers are not used for breeding, they are usually raised for meat. The beef you are used to eating is most likely from steers or heifers.

Now, the difference between the quality of meat from a steer and the meat from a bull is related to the animal's age.

Generally, both a young steer and a young bull (12 to 14 months of age) will offer you about the same quality of meat, which is good. But, as they get older, the quality of beef they produce drops off (which makes sense).

But the quality of beef from bulls actually depreciates faster than that of steers. So, beef from an older steer is more tender and probably juicier than that from an older bull because steers have lower levels of testosterone.

And one more thing! Because bulls are considerably bigger than steers, you'll get a lot more meat from bulls than from steers.

The Lifestyle of Bulls

Bulls are bred to be breeders. And when bulls are not working, they basically just lounge around eating and generally enjoying themselves.

It is important for bulls to eat well and rest when they are not "working" as they can lose significant weight when they are "working".

However, the first step to breeding is picking the right bull. This is a very important step because calves get a whopping 65% of their genes from their father. So be careful when you pick your bull.

Picking a Bull

There are two ways to pick a bull: raise your own bull or buy one from outfits that specialize in bulls.

Now, to run a cow-calf operation where there'll be a lot of cows and heifers, it is recommended that you lay off raising your own bulls and buy from a bull specialist or consider artificial insemination in part because of all that has been mentioned about the attitude of bulls.

But a more important reason is the possibility of inter-breeding. If you raise your own bulls and breed them in your practice, a bull might fertilize its sister or aunt or cousin or daughter. And interbreeding isn't a good idea because the calf could end up with a lot of medical complications.

One bull can service up to 25 cows throughout the breeding period (which lasts between 5 and 6 months).

So, if you have a small practice, you should be fine with just one bull. There's no need to over-saturate your farm with bulls and cause a nuisance.

That said, if you have cows of significantly varying sizes, it is to choose a regular or small-sized bull because allowing a big bull to fertilize a small cow can be dangerous for the cow and the calf at childbirth. If all your cows are big, a big bull will be fine.

Releasing the Bull

The next step is to let the bull loose among your cows. The bull knows his way around a mature cow and will walk up to 10 miles to show himself worthy of the cow.

However, calves are best born during the spring and fall because the weather is just right then. You can use the whole of spring and the whole of fall to allow your bull to fertilize the cows. This should give you a five-month period for the bull to do its work.

Retiring the Bull

Generally, bulls can breed when they are about nine or ten months and can keep going until they are 11 or 12 years old.

Realistically, your bull might need to be retired after five or six years of active service. Several issues can arise, such as structural problems (like a problem with the hooves that make it difficult for a bull to stand or a problem that prevents the penis from extending properly), which make it difficult for the bull to mate. Then there's also infertility, which makes it impossible for the bull to breed.

So, consider working around 5 of 6 years if you're drawing up a "breeding plan" and maximize that time period as much as possible.

When it is time for your bull to retire, you can either allow it to live its life out or slaughter it. The older the animal, the less the quality of meat it offers.

The Lifestyle of Steers

That you won't be breeding steers doesn't mean you shouldn't look out for those with good qualities.

If you are starting your own practice, prioritize the propensity for healthy weight gain and the production of tender beef. The reason you want to get the best bull is so your cows can produce the best calves.

So, if you are starting out by purchasing calves, look for the best. Ask to see the father of the calves to be sure.

Now, the life cycle of a steer can be seen in two basic stages: the growing stage and the finishing stage.

The Growing Stage

If you are starting your practice by buying cows and bulls and then breeding them, you'll start your steer rearing at the growing stage.

The growing stage is basically the time from birth to maturity when your steers are developing physically, mentally, and sexually.

If you are buying your steers as calves, buy them right after they've been weaned to avoid the complications of suddenly switching milk sources.

It is generally less expensive to buy steers as calves than as adults, but a lot more expensive to take care of them because you'll have them longer.

The same goes for breeding your own calves. You'll have to feed them and give them the shots that were mentioned in the previous chapter every year.

However, there are upsides to buying calves or breeding them. Most important, you know what kind of steers you want to breed, and since you've had them from birth you can tailor their food and general care to what you have in mind.

The growing stage (from birth to about 9 months of age) is a very sensitive stage. It is the period when a lot of the illnesses and parasites usually strike.

So, to get quality beef from your steers, pay attention to them during this stage.

The Finishing Stage

If you would rather not go through the stress of raising calves and are not ready to invest that much money in lifetime care, you can buy a full-grown steer.

It should be mentioned, though, that full-grown adults are a lot more expensive to buy than calves. However, they sure are a lot less expensive to rear than calves.

Steers that are at the finishing stage are basically just adult cattle that need to be fattened to bring in the money when slaughtered, but how effective this stage will be depends heavily on how effective the growing stage was.

Now, unlike bulls, steers do not have to "work". They just must eat, have proper vaccinations, and stay healthy so that when slaughtered, they'll bring in a lot of money.

Now that you can tell the various male folk in your herd and understand what those differences translate into, let's look at the various female folk in the herd.

Chapter 9: Cows and Heifers

In chapter one, we explained that not all cattle are cows. We also listed the different members of the herding community. In this chapter, just like we did in the last, we will be zeroing in on a pair of herd members: heifers and cows.

As we've seen in the first chapter, heifers and cows are both female, but heifers are female cattle that haven't had their first calf, while cows are female cattle who have had at least one calf. This is a very rudimentary explanation, as you can tell, and there's a lot more to know about heifers and cows. But first, let's find out the anatomical differences between cows and heifers.

Anatomical Differences between Cows and Heifers

Cows

Cows are mature female cattle, and the easiest way to spot them in a herd is to look between the hind legs. If there is an udder, then you're looking at a cow.

An udder is a pink sac-like organ that hangs down from the underside of a cow. The udder's four teats resemble cylindrical knobs from which milk is expelled. Usually, you'd almost always find a calf by the side of a cow, except where the calves have just been weaned off their mother's milk.

Now, to their physical appearance. Cows are usually smooth from head to tail. They have no prominent shoulder crests like bulls typically have, and their shoulders and hips are not as muscular as bulls.

Another way to tell if it's a cow is to look under the tail. Cows have a slit below their tail. This is the vulva, and it sits below the anus. It is from here that the cow urinates, mates with the bull during breeding, and pushes out the calves. Although both heifers and cows have vulvas, the vulva of a cow is much larger and much more defined than in heifers.

Heifers

When defining heifers more precisely, heifers are usually young female cattle that were born female (called heifer calves) and retained their female characteristics through adulthood. These two conditions must be fulfilled for a bovine to be considered a heifer, as there are cases where a calf is born female but grows up to develop secondary male characteristics. Such cattle are not called heifers, but as freemartins.

Most experienced cattle-rearers can easily tell a heifer from a cow just by looking at her. They notice the size and youth of the animal and can immediately tell. For an inexperienced eye, though, it's not as easy. Heifers are typically young cattle grown past the stage of being calves but still on the road to full maturity, which they usually hit by 3 or 4 years.

Anatomically, heifers do not have little hair, a sheath, or a sac between their legs like steers and bulls. They do have udders, but the udders are almost absent, and the teats are nearly impossible to see, even between the hind legs.

Just like cows, heifers have a vulva under their tail, below their anus. It's not as pronounced or as large as a mature cow, though. By the time the heifer is bred and is about to calve, the vulva and udder increase in size, resembling what is seen in more mature cows. The udder still isn't as large as a mature cow until the heifer has calved.

Heifers that have never calved by the time they are older than two years of age are called heiferettes, while a heifer carrying her first calf is called a bred heifer.

A Word About Raising Cows

You can raise a cow for either of two purposes: for its beef or for its milk. Whatever path you choose, your choice will affect how you raise your cattle. Allow us to chip in a word of advice. For a small scale backyard herdsman, raising a cow for her milk is not a wise investment.

Usually after a cow has had her baby and can produce milk, she will keep lactating if you keep milking her. You can milk a cow for about two years before the udder finally dries up completely, even if she doesn't have another calf during this time. Problems can arise when trying to sell the milk because of numerous rules and regulations surrounding dairy production.

For instance, in 13 states, you're free to sell raw milk in a retail store. In 17 others, you can sell raw milk on your premises, and 8 states only allow you to sell milk through cow-share agreements. A cow share agreement is when cow owners are paid money to board, feed, and then milk their cattle. Still, in 20 states, it remains illegal to sell raw milk straight from the farm and unpasteurized. In these states, you can only milk your cow for personal use.

Now, before you say that doesn't sound like a bad idea, consider this. Maintaining a milking cow is not cheap, and if you're not selling the milk, raising a cow to make dairy for subsistence use only might not make economic sense as you spend more than you save.

True, your cow can give you excess milk, much more than you can consume, and you won't have to buy milk. But how much are you really saving? A gallon of milk is about $3.00. On average, you're buying about two gallons a month, right? That's $6.00. How much are you really saving?

Perhaps, to supplement that, you can process your milk into cheese and butter, so you get to save on those. But on a small-scale, raising cows for their milk still has more drawbacks than benefits.

This is why it's best to raise cows for their beef on a small scale. Like all cows, beef cows will still produce milk, although at much lower quantities, and that will suffice for your family's needs.

A Word About Raising Heifers — Choosing Replacement Heifers

In a cow/calf operation, everything rises and falls on the selection of replacement heifers. Your female cattle are the future of your herd. If they aren't selected with careful thought, it can be bad news for you.

The first thing you want to consider is weaning weight because puberty and weight are two closely related factors with female cattle. It's best to set apart the heaviest and the lightest calves, think upper 1% and lowest 25%, respectively. While you want heavy heifers, heifers too heavy might be too big for your environment, especially if you're a small-scale farmer.

Now, when selecting, make sure you're choosing based on their actual weaning weight. It's important to do this because you are going to develop their feeding program based on their weaning weight. This will help them reach puberty (about two-thirds of their adult weight) on schedule.

Other factors you want to look into before selecting a replacement heifer is the conformation of its body. Check out her feet, her legs, and her body type. Also, check out her disposition. Don't forget to meet her dam as that gives you a picture of what the heifer is most likely going to look like by the time she becomes a dam herself.

The Relationship Between Feeding Heifers and Calving Time

Now, what you're about to read may sound super weird and unscientific, but it is tested and proven. To prevent a cow from calving at night, the most practical and the easiest way to make that happen is to feed your cows at night. Experts can't explain the science behind it, but they think hormones might be involved.

Research has been conducted to study the motility of the rumen. From this study, as calving time approaches, rumen contractions are reduced. The fall in contractions begins about two weeks before calving and then falls more rapidly during calving. How does this relate to feeding at night? Well, links have been drawn between nighttime feeding and the rise of intraluminal pressure at night with a decline during the day.

Several studies have proven this phenomenon, but we will focus on one for our discussion today. In Iowa, there were 1331 cows from 15 farms. These cows were fed only once daily and only at dusk. When it was calving time, 85% had their babies between 6.00 am and 6.00 pm. It didn't matter if the cows started the nighttime feeding program a week before calving or two to three weeks before calving time. Most had their calves during the day!

Now, while achieving nighttime feeding for a large herd on a large ranch might be difficult and would require a more sophisticated process, it's easier for smaller farms. Large ranches have it a little tougher. One way to make it easier for large ranches is for managers to feed the cows earlier in the day and leave the nighttime feeding to heifers with their first calves. You want to give priority to the heifers as they require the closest observation during the calving season. It's their first time, remember?

The Twin Problem

You know the deal with expecting twin calves. If they are different genders, the heifer (freemartin, more correctly) is affected by her twin brother's male hormones. This makes calving unforeseeable in her future, but when the twins are both heifers, there's no testosterone interference. Hence, both heifers should come out just fine with their reproductive abilities intact.

Replacement Heifers: To Buy or To Raise

The decision between buying and raising your replacement heifer is something many people have attempted to help cattle farmers with, but there remains no one answer. There's only the best answer for you. Here are a few factors you want to consider when deciding between buying or raising your replacement heifer.

Herd Size

How does herd size affect your choice between buying a replacement heifer and raising one? And which is the more economically smart thing to do? To raise these heifers or to buy them?

For small-scale herdsmen, buying the replacement heifers might be more cost efficient than raising them due to economies of scale. But larger-scale farmers might find it more economical to raise heifers.

But even large-scale farmers still prefer to buy their replacement heifers rather than raise them. This frees up resources and time, which they can channel into other more pressing areas on their farm.

Facilities and Pasture

Heifers are more demanding to manage than cows, both financially and otherwise. You must consider this too when deciding to buy or raise replacement heifers.

Heifers need to be managed on their own away from the other members of the herd if they are to reach their peak maturity level for breeding. And you need to begin this separate management when the heifer calf is weaned, especially within the first two to three weeks of weaning. During this period, your heifer calf is very vulnerable to illness; hence you must give her extra special attention. If you don't develop your heifers carefully, they won't hit puberty and be ready to breed on schedule, which generally should be when they are between 14 to 15 months old.

Another aspect of raising heifers is feeding. Growing heifers' nutritional needs are different from the nutritional needs of other members of the herd. To wean and develop your replacement heifers properly, you need to provide more pasture. You must get a secure holding pen to protect the heifers from the bull before its breeding season.

Considering all we have just mentioned, it's easy to see that managing heifers is tough, and there are no shortcuts. Taking shortcuts when developing your heifers will only affect their productivity in the long run. But if you buy your replacement heifer instead, you provide more pasture for about 10% more cattle.

Can You Afford To Raise More Heifers Than Needed?

If you raise your replacement heifers, remember that you can't just raise the exact number of heifers you need because not all of them will stay healthy. A few of your heifers might have to be culled for several reasons ranging from poor structure to poor weight gain.

If you raise your replacement heifer, consider raising at least 45% more heifers than you need. It's going to cost you more and tie up your capital. Best-case scenario, it will be at least one year before you can sell the heifers you don't need and make your money back.

Herd Health

Despite the difficulty involved in raising replacement heifers, many farmers still raise their own due to health concerns. If you're buying your replacement heifers, you're not sure where these heifers are coming from or to what they've been exposed. You have only the seller's word. There's always going to be a risk of introducing a foreign disease into your herd. A sickly herd is a huge problem you want to avoid. And if we are following the highest level of biosecurity, then you want to maintain a closed herd which means you should raise your own replacement heifer.

But if you'd prefer to buy the animal, then take these steps:

• Ensure you only buy heifers from a reliable source with a clean bill of health. If you're not sure what to look out for, meet with your local vet to give you the health criteria the heifer should meet.

• Always quarantine newly bought animals.

• Always follow through on your vaccination program.

Genetic Base

The demand for high-quality beef is increasing, and with beef, quality rises and falls on genetics. The genetics of a cow can affect the profitability of your herd for over ten years - up to 14 years, in fact!

This is one area where raising your replacement heifers trumps buying. As a producer, you can select cattle based on specific performance, carcass, or maternal traits to sire your replacement heifers.

Also, and even more important, if you're raising your replacement heifers, you get to select the heavier calves born within the first 60 days in the calving season. Such heifers have a higher chance of hitting their optimal weight by the onset of puberty. Plus, these heifers usually come from the most fertile cows able to conceive in the earliest days of the breeding season. And if there are heifers that fail to conceive, raising your replacement heifers means you get to cull them.

Now, does this mean you can't select fertile females through buying? No, there are many reliable sources from which you buy get good heifers. Just look out for sources that place a premium on strict selection and quality genetics.

If you want to quickly improve your herd's genetics, it might be a great idea to choose your heifer from outside sources. Selecting from outside sources is also good if your gene selection is limited because of heavy culling either due to age or drought.

Calving Difficulty

There was a study conducted by the Colorado State University and the University of Nebraska Meat Animal Research Center. According to these studies, first-calvers at two years have calving difficulties compared to mature cows at the age of three. This condition is known as dystocia.

Dystocia has two main causes: the small size of the pelvis in immature heifers and the heavy birth weight of calves. Pelvic size cannot be fixed, but something can be done about calves and their birth weight.

Heavy birth weight is usually caused by the sire's genetics. Hence, to reduce it, you can breed your heifer with a low-birth-weight sire or a calving-ease sire. This is an advantage you only get if you raise your heifer.

If you're buying, you might not be able to confirm that the dam was bred with a calving-ease bull, but you can mitigate this by buying your heifer from a trusted supplier.

Now, remember that using a calving-ease bull does not necessarily mean that calving season will be dystocia-free for your heifers. Remember that pelvic size is another contributing factor. So, if the heifer isn't fully mature by the calving season, she could still have dystocia.

Other factors, such as being a first-time calver or incorrect presentation of the calf, can also make your bred heifer experience dystocia.

So, keep these in mind and consider your ability as a producer to handle these issues should they arise. If you can't, it might be better to buy a replacement heifer.

That said, let's look at the advantages of each option for replacing heifers in your herd.

Benefits of Raising Replacement Heifers

Greater Genetic Control

Let's assume that your breeding program already involves a couple of generations specifically selected for maternal traits such as milk production, calving ease, fertility, maternal instinct, and stay-ability. In such a case, getting a replacement heifer from elsewhere would be extremely difficult.

Also, finding heifers with the matching genetic profile suited to the environment that maximizes longevity can be quite difficult.

Greater Control Over Herd Health

If you operate a closed herd system, it's easier to minimize diseases within your herd. Diseases such as bovine viral diarrhea, venereal diseases, and respiratory diseases are more easily controlled when you develop your own replacement heifers on site.

Benefits of Buying Replacement Heifers

It Frees Up Your Resources

When you buy your replacement heifers, you buy only as many as you need. In contrast, when raising them, you wind up with more than you need because, in the end, most end up calving. This consumes extra pasture, facilities, space, and feed, which could have been channeled into raising cows that would calve in the end.

Takes Less Time to Expand Your Herd or Switch a Breeding Program

If you expand your pasture ground or get access to more inexpensive feed, you might increase your herd size. To do that quickly, your best bet is to buy from an external source, as raising new heifers would be time consuming.

It's also possible that a new marketing window involving a different genetic sub-population opens up, and you wish to explore it. Whatever the case, buying from an external source is the fastest way to take advantage of that opportunity.

Might Be the Only Way to Get Superior Heifers

If you buy from a replacement heifer specialist, you could end up with a more superior heifer than you'd have been able to produce on your own. When buying from a specialist, you can specify the genetic profile of the composite and purebred heifers, the breed cross, and the sire to which the producer breeds the heifer.

Most commercial developers use artificial insemination along with estrous synchronization to increase the genetic merit of the resulting calves and eliminate any chance of transmitting reproductive disease. This technique also makes it possible to develop heifers that will conceive and calf over a shortened time frame.

May Be the More Affordable Option

It can be quite costly to raise heifers on your own, especially if you don't have access to cheap feed resources. Plus, if you don't grow your heifers fast enough from the weaning stage to the breeding stage several bad things can happen including delayed puberty, low conception rates, an extended calving season, and increased cost to maintain each pregnant heifer. Now, if calving season is extended and most of your heifers are calving late, the weaning weight of your calves could go down, which affects profitability.

Chapter 10: Cattle Breeding and Reproduction

When it comes to beef production for commercial purposes, no aspect is as important as reproductive efficiency. It doesn't matter whether you crossbreed, whether your cattle have superior genetics or whether you've been managing your cattle well. If reproductive efficiency is only 50%, it will affect your business dramatically. This is why you must learn everything you can about preparing your cattle to have excellent reproductive efficiency.

When you manage your cattle properly, you could have a calf crop that's higher than 90%. The very least you must cover your production expenses is a calf crop of 85%. Anything lower than 75%, and you'd record major losses. A good goal to work toward is a calf crop of 95% within a calving season of 60 days and an average weaning weight of 500 pounds.

How to Prepare Your Cow to Conceive Successfully

For a healthy pregnancy and a healthy baby calf, your cow will need a lot of tender loving care. Ensure that she's in great shape physically. Also ensure that she's given the proper care for her age and provided with all the preventative healthcare measures to make breeding successful. Here are a few pointers to help you do this.

1. Assess Your Cow's Body Condition Scores

An underfed pregnant bovine is prone to several negative consequences, including:

Malnourishment: Since most of the already insufficient nutrients are directed to the fetus, the cow ends up malnourished. Malnourishment makes it difficult for the dam to give birth.

Poor Colostrum Quality: Poor feeding affects the quality of colostrum the dam produces. This means the calf receives fewer antibodies, which affects the strength of their immune strength.

Decreased Milk Production: If cows are not well fed, they don't produce as much milk.

Slower Rebreeding Rate: When pregnant cows are underfed, they longer time to rebreed. This means they won't be producing a calf yearly as they should.

Given how important nutrition is to the reproductive efficiency of cows, farmers have come up with a numerical system to assess a cow's body condition. This numerical system is called Body Condition Scores (BCS).

Body Condition: Scores refer to a set of numbers that indicate the body condition (relative plumpness) of a cow before pregnancy, and during and after. The thinner the cow, the lower the BCS score; fatter cows will have a higher BCS score.

Here are the 9 points and what they indicate:

1. Means the cow is so weak and emaciated, you can see the bones distinctly. This is a very rare score, and when it occurs, it is typical because the cow is diseased or riddled with parasites.

2. Means the cow is very emaciated to where you can pick out the ribs distinctly. Muscles in the hindquarter and shoulder are also typically atrophied. Nonetheless, such cows are extremely weak.

3. Means the cow is very weak, and there's no fat in the brisket or overlaying the ribs. Cows with a BCS of 3 also have visible backbones, and there's a reduction in the muscles of the hindquarters.

4. Means the cow is only slightly thin with only a few visible ribs (3 to 5) and a visible spine but there's no muscle depletion, and you can see fat on the hips and over the ribs.

5. Means the cow is in moderate condition. So, there's no backbone jutting out, the spine looks smooth, and most ribs are covered with fat except the last two.

6. A cow with a body condition score of 6 is well-conditioned. Such cows usually look smooth all over their bodies with rounded backs. You can find fat covering the tail-head, in the brisket, over the pin bones, and over the ribs.

7. A body condition score of 7 means that the cow has enough flesh, and the brisket is fatty, as is the tail-head. The body has an overall soft and rounded appearance with smooth ribs and only slightly visible hip bones.

8. This means the cow is over-conditioned. Because of the excess fat, the cow's neck looks thick and short, and the back has a squarish appearance. The entire bone structure is covered by fat.

9. A body condition score of 9 is too high and usually means the cow is very obese. Thankfully, just like a score of 1, a score of 9 is quite rare.

So, what's the best score for your cows to ensure a healthy pregnancy and calving? Well, generally, anything between 5 and 6 is fine but for a first-calf heifer, a minimum body condition of 6 is desirable and if they are calving in the winter, then aim for a higher BCS.

To ensure that your cow attains the perfect BCS for calving, you must evaluate her at specific times. So, evaluate your cow once she enters the second trimester, right before she calves, and before she breeds. This way, you can catch any deficiencies in time and make the necessary adjustments to bring your cow to the perfect body condition.

2. Pay Particular Attention to the Younger and Older Cows

Pregnancy takes a toll on all cows but not as much as in first-calf heifers, younger cows ($\leq$3 years), and older cows ($\geq$9 years). You need to pay extra attention to these age groups.

Caring for First-Calf Heifers

A first-calf heifer is still in the process of maturing even as she carries her calf. So, her body is undergoing a lot of stress and, therefore, needs all the help she can get.

If you did a good job selecting or developing your replacement heifer and ensured they were well nourished, your heifer should have successful calving by the time she's two years old. She should also be ready to rebreed in a year and wean her heavy calves. A well-nurtured first-calf heifer becomes a more profitable animal in the long run.

Here are a few tips to help you support your first calf heifer:

- Be Strict With Selecting Your Replacement Heifers

We already showed you how best to go about this in the previous chapter. Kindly go through it again if you need to.

- Feed Stage-Appropriate Feed and Rations

Weaning to Breeding: Your heifers need to gain roughly 1.25 pounds daily. Now, for one reason or another, your heifer might be falling behind and need to catch up to reach the target weight of 65% mature weight. Even in that case, try not to exceed a daily weight gain of two pounds. If your heifer needs to put on over two pounds daily to attain her target weight for breeding, she's probably not a great candidate for breeding.

To achieve a daily weight gain of 1.25 pounds, your heifer would need to eat about 12 to 15 pounds of the dry matter daily. This can be just pasture, provided you're sure that it is high in protein and energy. If not, then supplement pasture with concentrate.

Breeding to Calving: The daily weight gain target for pregnant heifers is 0.8 pounds. So, they need to be supplied with about 20 pounds of the dry matter daily. Remember, they are now sharing with their baby calves, and that's why they need more food to put on less weight.

Now, remember that as heifers approach calving, it's tough to get them to gain weight, and that gets tougher after calving. One way to work around this is to supplement her daily rations with concentrate to ensure she's at a BCS of 6 or 7 before calving.

After Calving: A heifer does not attain full maturity until her second calving, especially heifers of larger breeds that mature late. Even after calving, your heifer is still growing, and you want to keep her at a BCS of 5 or 6. They could make do with the same rations given to other members of the herd. Often, they might be better off continuing to be managed with the first-calf heifers.

- **Breed First-Calf Heifers First**

A good breeding schedule tip is to breed first-calf heifers for two to three weeks before the cow is in the herd because you want the first-calf heifers and the more mature cows to remain in sync if you have to rebreed them all at the same time.

- **Select the Right Sire**

When picking out the bull to breed with the heifers and cows, select one with convenient birth weight. The bull should also come with a calving-ease expected progeny difference (EPD). Both factors tell you how easily the calves from that bull were born. They are also mutually exclusive as the weaning weight of the bull affects his EPD along with other factors like calf body shape. Generally, larger calves tend to cause dystocia.

Caring for Older Cows

As your cow ages, understandably, her productivity will also decrease. Conception rates will fall, her calves will have a lighter weaning weight, and her ability to forage will be less (probably due to old age).

So, here's how you can support your older cows. If their udders still function such that they can nourish a calf to a desirable weaning weight, it might be a great idea to either manage them with the first-calf heifers or with the more mature 3-year-old cows.

In either of these places, the competition is less, and your senior cow can get all the nourishment she needs.

3. Don't Forget the Reproductive Vaccines

Reproductive diseases can negatively affect the health of your cow and the profitability of the herd. They make breeding impossible, abort calves, and tamper with growing and milky.

Here is a vaccination schedule to help you:

Brucellosis: Should be administered between 4 months old to 1 year. If you intend to include a heifer in your breeding stock, make sure she receives this vaccination within the aforementioned time frame.

Leptospirosis: Cows and heifers should be vaccinated for leptospirosis yearly, at least. If there's been a case of leptospirosis in your herd before, then vaccination might need to be more frequent; twice a year.

Vibriosis: Before breeding, your heifer should get this shot twice. Mature cows need booster shots yearly as well.

Trichomoniasis: This vaccination is super important, especially if you're on the west side of the US, as trichomoniasis is more common in that part of the world. Confirm with your vet whether your heifers and cows need this vaccine.

Bovine Virus Diarrhea (BVD): BVD isn't just a reproductive disease. It also affects the immune, respiratory, and digestive systems. The BVD virus comes in two strains. So, make sure that your vaccination programs cover your cattle from the strain they are likely to encounter.

Infectious Bovine Rhinotracheitis (IBR): This does not just attack the reproductive system of cows; it also affects the eyes and trachea. This shot should be administered annually for heifers and cows.

Always ensure that whatever vaccine you're giving your bred cows/heifers is safe to use in pregnant animals.

4. Scheduling the Due Date

It's important to have a rough estimate of when your cow is to calve. It helps you care for your cow appropriately at different stages in her pregnancy. Gestation (pregnancy) in cows takes roughly 285 days.

Now, since we're here, how do you confirm that your cow is pregnant? Well, the common method is by palpation. During palpation, the expert inserts a gloved hand into the animal's rectum and feels the reproductive organs for signs of pregnancy.

For highly experienced experts, pregnancy (or the absence of it) can be confirmed 30 days after breeding, but this requires the experience of an expert hand. Generally, 45 days is the more commonly used time frame. In 45 days after a heifer or cow has been bred, pregnancy (or the lack of it) can be accurately confirmed.

Choosing the Right Bull for Breeding

In the first chapter of this book, we looked at the fact that you should have a goal when going into the beef cattle raising business. If you have a goal for your business, then you can make the best choices. The bull you select should be one able to complement your heifer/cow in her areas of weaknesses.

Here are a few factors you want to consider:

1. Is the Bull to be used on a Heifer?

If yes, the top factors to consider are birth weight and calving ease.

2. Are You Going to Retain all His Heifer-Calves?

If you intend to retain all his daughters, then choose a bull with a track record of producing cows with excellent maternal instincts. You're looking for traits like fertility, udder conformation, milking ability, as well as mothering ability.

3. Are You Going to Sell the Offspring of the Bull as Feeder Calves?

If you intend to sell the offspring from this bull as feeder calves, the weaning weight of the calves should be one of the most important factors to consider.

4. Are You Going to Sell the Offspring of the Bull as Beef?

If you're selling directly to consumers as beef, then you want a bull known for its fantastic carcass merit.

5. Reproductive Soundness

Conduct a breeding soundness exam before bringing your bull to the heifer/cow to confirm that he is fertile.

6. Structural Fitness

The body structure is also important for reproductive efficiency. You want bulls that move confidently, are strong enough to mount cows without tiring easily, are not diseased or injured, have no swollen leg or joints, have good vision, and that his mouth and teeth are in excellent condition.

7. Look at Your Bull

Carefully observe the bull for signs easy to miss. Keep an eye out for muscling, disposition, color, and the body condition score. For the start of the breeding season, a BCS of 5 or 6 is best.

8. Assess the Bull's Performance

Carefully consider the bull's own performance in key areas such as weaning and yearly weight because that tells you the offspring that would result. More important, learn about the bull's EPD because this goes a long way in determining how easily the mother births her baby.

How to Care for Your Bull

Begin caring for your bull right after weaning. Once he's been weaned, he needs to hit about 2.5 pounds daily to grow and mature properly, and bulls continue to grow well into their third year, attaining a 1,000-pound weight from 600 pounds weaning weight.

So, ensure your bull has access to food continuously (about 23 pounds of dry matter daily) and has a body condition score of 6. If it's an older bull, then his diet should consist of about 25 pounds to 31 pounds of dry matter (depending on his size) to maintain his weight.

Now, here's an important tip: You should never leave your bull with the cows all year long. Have only your bull with the cows during breeding season and breeding season alone. Typically, this season lasts between 60 and 90 days.

There are several advantages to this. One is that all members of your herd would be at the same production stage (pregnancy, lactation, and rebreeding). Also, you'd have a tight calf crop. All your calves would be about the same age. This allows you to care for your herd and calves more precisely and easily when they all have the same health and dietary needs at roughly the same time.

Leaving the bull with the females all year long is risky because bulls are randy animals. Sometimes, certain heifers attain sexual maturity ahead of the expected time. The bull can prematurely breed such heifers, and that could make things difficult for you.

A bull should be kept in a clean, dry area protected from the elements but also has enough space for him to pasture and exercise. You can put the water trough and feed bunk at opposite ends of the pasture so that he's forced to move around and exercise. Remember, he needs to strengthen his muscles and bones.

Now, don't forget to vaccinate your bull just as you vaccinate the cows. The mature cows and your bull could share the same schedule for vaccinations. Here are the vaccines your bull will need:

Reproductive vaccines, including vibriosis, leptospirosis, and trichomoniasis (possibly).

Respiratory vaccines for bovine virus diarrhea (BVD), bovine respiratory syncytial virus (BRSV), infectious bovine rhinotracheitis (IBR), and parainfluenza-3 (PI-3).

Deworming and aggressive fly-control are also very important for bulls to keep them healthy.

Going the Artificial Insemination (AI) Route

It's possible to get your cows and heifers pregnant without breeding them with a bull. This process is known as artificial insemination. You're probably familiar with the term. It's where you collect semen from a bull and use it for breeding your cow without bringing the bull to mate her naturally.

Artificial insemination is very popular in the dairy industry in the United States, with about 2 out of 3 dairy cows being bred by artificial insemination. In the beef industry it only about 5% to 10% of beef cows are bred using artificial insemination.

Now, let's examine the good and bad of artificial insemination and see if it's right for you.

Pros of Artificial Insemination

1. High-Quality Sires

This is the biggest advantage of artificial insemination: access to top-tier AI (Artificial Insemination) sires. Such bulls typically have a proven track record and produce offspring with heavy weaning weights, high carcass merit, and excellent replacement heifers.

2. No Need for a Bull

There's no hiding the truth. Having a bull around can be tough for most herdsmen. So, not having to deal with one would be a welcome relief for most, especially for small-scale rearers.

3. Might Be the More Affordable Option

Artificial insemination might be less expensive than natural service, especially if you have a small herd with only a few cows.

Cons of Artificial Insemination

1. Conception Rates are not as High

Natural service will always have higher conception rates than artificial insemination. If, after two attempts with artificial insemination, your cow does not become pregnant, it might be advisable to go the natural route.

2. AI Requires Labor, Skill, and Equipment

In natural service, the bull does all the work. For AI, it's different. You're going to have to put in the time, equipment, and effort. You might have to go with an AI technician to increase chances of success and to make things easier for you, but it is also possible to do it yourself with the right education.

Now that you know all about cattle reproduction, let's get into the more practical aspect of calving.

Chapter 11: Calving and Caring For Newborns

Now that you know how to get your cows pregnant, it becomes imperative to know how to care for them in pregnancy and post-pregnancy. You also need to know how to take care of the new calves. And that is what this chapter is about.

But before we go into all of that, here are a few quick facts about pregnant cows:

> • Many people peg the gestation period for cows at 283 days, but it could be from 279 to 287 days. This variation could be because of the gender of the calf. Cows carrying bulls sometimes have a longer gestation period than cows carrying heifers.

> • On average, a cow can get pregnant 55 days after she has calved, but this can take up to 10 days longer if the said cow had calving difficulties or is a first-time mother. Cows that fall ill and lose weight after calving could also take longer before they could get pregnant again.

> • Under normal circumstances, your cow should be able to give birth to one calf every year if she is bred.

With all that said, let's look at what you can do to keep your pregnant cows and heifers healthy.

Feeding Your Pregnant Cows and Heifers

The goal of proper nutrition with pregnant cows and heifers is to ensure that they remain healthy throughout their pregnancy, deliver healthy calves, lactate well, start their next cycle promptly, and then be healthy and ready when the new breeding starts.

So, it is clear to see that nutrition for your pregnant cows and heifers is something to pay proper attention to.

With that said, the first thing to note is that at the very early stage of pregnancy, their feed need not be changed. What you were feeding them at the breeding stage is what you'll continue to feed them throughout the early stage.

But as the fetus inside them continues to develop, their nutritional needs will continue to increase. This basically means that you must eventually feed them like they are eating for two because they will be.

Two months to when your cows will deliver is when a lot of the fetal developments happen. So, this is the period when you want to improve their nutrition and feed them like they are eating for two.

Now, your younger cows and heifers will need even more protein and better nutrition than the older cows because the younger ones are still growing while still pregnant, but the older cows are no longer growing.

As you increase the foliage you give your pregnant cows, there must be a concurrent increase in the protein they eat.

The reason for this is that your cows and heifers need protein to properly digest the foliage. Also, protein creates the right environment for rumen microbes to grow.

These rumen microbes help the cows extract the energy value from the foliage they are eating. Don't forget that the energy extracted would be really needed, especially when they are pushing.

They also need the foliage to be properly broken down in the rumen so they can maintain a healthy weight throughout their pregnancy and even post-pregnancy. Protein can be instrumental to keep them at a healthy weight.

So, what you want to figure out is how to increase the protein content of your pregnant cow's food. And remember to give the younger cows more than what you give the older cows.

You could increase the protein content by increasing the protein source food in your cows' diet or by adding protein supplements to their meal.

The general rule of thumb is to increase the protein content of your older cows to about 7 to 8 percent while you increase that of the younger cows and heifers to about 8 or 9 percent.

Besides protein, you also want to make sure that your pregnant cows are getting enough vitamins and minerals. Calcium and phosphorus are other nutrients to pay attention to. Consult with your vet to tailor a meal plan for your pregnant cows and heifers.

Other Things to Pay Attention To

- It is *absolutely important* for your pregnant cows and heifers to get as much exercise as possible. 30 minutes of moderate exercise twice a day should work just fine.

- You could consider massaging your cows' udders for a couple of minutes every day to help increase circulation.

- You'll need a maternity/calving pen where your cow will be delivering her calf. If you can afford it, have a calving pen in a separate barn. But if you can't, just make sure it is as far away from where the other cows are as possible in order not to agitate them.

- Also, make sure this maternity pen is close to a handling facility (headgate and squeeze chute) in case if you need to assist with the delivery.

- You do not want to enter the calving season unprepared. So, make sure that you have a plan for it. Plan your schedule so you always have someone around and consult with your vet to come up with precautions that everyone in the family or practice must take.

- You need to prepare your calving kit. Your calving kit should contain obstetrical sleeves (preferably disposable ones), lubricant (non-detergent soap works well), antiseptic (preferably hypo-allergenic), obstetrical chains (30 and/or 60-inch chains), mechanical calf pullers, and injectable antibiotics.

Now that we are on the topic of calving let's look at what signs tell you your cow is ready to deliver.

Signs of Calving

- You'll notice that your cow's udders are really full.

- Her birth canal will look really long and squishy, showing you it is preparing for a calf to come out.

- When it's really close to the time, you'll see discharge from the birth canal.

- You'll notice her mood begins to change. She'll be cranky and anxious.

Preparing for Calving

As your cow approaches her final month of pregnancy, you need to move her into the maternity pen or space to acquaint her with the new area.

And as you notice the signs already mentioned, you need to make sure that the calving space is calf safe. Remove everything that could pose harm to the babies. Also, make sure that the space is clean and dry. Have your vet's number on speed dial, just in case things don't go as planned.

Calving

- As your cow or heifer enters labor, you'll be able to see the fetus in the birth canal.

- Cervical dilation and contractions start. This should last between 4 and 8 hours. If it goes on for longer than that, call your vet.

- Next, her water breaks, and she goes into active labor, which is marked by straining. The time from when her water breaks to when the calf drops should be between 2 and 4 hours, but for a first-time mother, active labor should last for about 60 to 90 minutes Active labor for older cows is 30 to 60 minutes. If it continues beyond that time without the calf dropping, you need to call your vet. You might have to step in but your vet will tell you what to do.

- After the calf has dropped, your cow should pass the placenta. If this has not happened within 12 hours, it means your cow has retained the placenta, and you need to call your vet.

Calf Handling

After the calf has been born, it is time for you to step in, make sure that the calf is alive and then take care of it. But before you get to the calf, be careful of the new mama, as she might not take kindly to you touching her baby. So, do not sneak up on her; make sure that she is OK with you taking away her baby. Also, try not to go through the calf handling process alone in case you need help with the mama.

With that said, here are the things you need to do:

- First off, make sure that the calf is breathing. If it is not breathing, try cleaning the calf's nostrils and mouth with wet wipes. The nostrils might be blocked by mucus. You could also try encouraging the calf to breathe by vigorously rubbing its back or tickling its nose with a piece of straw.

- Do not hold a newborn calf upside down as that could squish its internal organs unto its lungs, preventing it from breathing properly.

- Once you've ascertained that the calf is breathing, examine its general wellbeing. The calf should be able to move and the body should be warm within five minutes. It should attempt to stand within fifteen minutes and actually stand on its own within an hour. If it can't meet these milestones, call your vet.

Calf Nursing

Once you've ascertained that your calf is breathing, the next thing is to ensure that it gets as much colostrum as possible.

When calves are born, they do not have a very good immune system. Colostrum should help them strengthen their immune system. Colostrum is the first milk that mammals produce, including humans. And as the hour passes, the amount of colostrum they produce decreases.

So, if you want your calf to get enough colostrum, ensure that they nurse no more than 30 minutes after they are born. If the calf can't nurse within 30 minutes, you must bottle feed it colostrum.

The colostrum must have been frozen, and then before you feed, slowly thaw it, then feed the calf. The amount of colostrum you feed the calf should be 5 to 6 percent of the calf's body weight. Make sure that the calf is fed colostrum within the first six hours of its life and then 12 hours after it was born.

If the calf is too weak to be fed by mouth, you might have to resort to a stomach tube, but if you encounter this kind of issue, consult with your vet first.

Calf Health

After your calf has nursed for the first time, it is time to disinfect her navel. Use a 7% iodine solution in a container and then dip the calf's umbilical cord and navel into the solution.

You want to dip it as opposed to spraying because when spraying it is easy to miss a few spots. If you've had a history of navel infections on your farm, consider doing the dip again after 12 hours, just to be on the safe side.

Moving on, if you notice things like rapid breathing, dry muzzle, abnormal posture, lowered head and ears, call your vet because that is not normal.

Calf Identification

The next thing to do is to ID your calf so you can remember when each calf was born and their parents, but if you have just one calf, you might not need this step.

With identification, you could use physical hanging ear tags, radiofrequency ear tags, or tattoos. Whichever one you use, the ID should be a combination of the year they were born and a number that represents the order in which they were born. It is generally accepted to denote the years as letters thus: H=2020, J=2021, K=2022, L=2023, etc. The letters I, O, U and U are not used.

Therefore, the fourth calf to be born in your practice in 2020 would typically have H4, H04, or H004 as its ID, depending on how many cows you have.

One more thing: IDs are usually attached to the ear. There are two trains of thought when determining which ear. Many people append the ID on the left ear for all their calves because it is easier to see it as the cattle go through the handling facilities. There are those that attach the ID on different ears for different genders. So, if they attach the ID on a bull-calf's right ear, they'll attach the ID on a heifer calf's left ear. This allows them to identify the calf's gender at a glance.

After you've ID'ed the calf, you want to make records. Record the date and time of birth and the mama and papa. You also want to record the calf's weight, which you should take within the first 24 hours of its birth.

Calf Castration and Implants

This only applies to bull-calves. Castration and implants need not be done that day, especially considering that you need to observe the calf to determine to use it for breeding.

Castration is the removal of the testes, which makes the bull-calf a steer. But a growth implant may be implanted into the steers to make them grow almost as big as a bull. Do consult with your vet, who should help you determine if a growth implant is a good idea for your steer.

Finally, if you castrate and/or insert the growth implant, try to do it before you wean the calf.

Calving might seem scary to you, but it isn't. Mostly, your cow can do the calving herself, and if she needs help, you and your vet can assist her.

Chapter 12: Expert Tips for Your Beef Cattle Business

It would be easy to think that having all that information on how to set up a proper cattle rearing outfit would guarantee success. However, it takes more than knowing how to run a successful outfit. You need to know how to run a successful business, and this chapter is where we attend to that.

A Quick Guide for Beginners

Costs

Setting up a cattle rearing outfit will cost you a lot of money upfront, regardless of the size of the practice you want to run. You'll be spending a lot on everything from the land (if you don't have already), setting up the pasture, erecting fences, and setting up the facilities and equipment that have been mentioned earlier in this book. And this doesn't even include buying the cattle that make up the practice.

Plan properly for it. Having a sustainable source of income before starting the outfit would be a good idea. Still, often it's impossible to get started without a loan. Consider using the services of a financial adviser to better understand your options.

You also want to draw up a budget for your practice. Online budgeting tools and land grant university programs can be found free of charge; take advantage of such options! Also, consider the savings afforded in buying used farm equipment. It not only saves you money, but it is also good for the environment. *(Do try to new feeders, though, as mentioned earlier.)*

How Much Land?

The minimum amount of land you should aim for is ten acres; with that much land, you should be able to run a small cattle-raising operation. For a standard outfit, start with at least 30 acres. To start small – but are looking to expand later – buy land in an area with prospects where you are sure you can buy more land later, avoiding having to move your practice down the road.

The Easier Practice

If you are clear on what kind of practice you want to run (that is, feeder or cow-calf), then by all means, do what you want to do. But if you haven't decided yet (or you *have,* but are open to suggestions), you'll want to hear this. As a beginner, you should start out as a feeder practice, and there are good reasons for that.

One, a feeder practice is more affordable to start than a cow-calf because with a feeder practice, you can go right ahead and buy a mature cow and earn money almost immediately. Mature cows are more expensive to buy than babies, but they are also less expensive to care for than babies, seeing as you'll have them for just a few years.

On the issue of cost, with a cow-calf practice, you must set up different facilities for cows and calves while you can use one-size-fits-all facilities for feeders. Furthermore, feeder practice is less stressful to run. This is especially good news for someone with no experience.

You won't have calves who are still delicate and require a lot of attention.

Something else to think about is that a feeder practice offers you more opportunity to experiment. To start with two cattle, you can buy two breeds and decide which works best for you. Plus, you won't be stuck for years with a breed you think doesn't work. Still, you may be stuck with a calf till it grows old enough to be sold, unless you want to sell it as a calf and take a loss.

Breeding

If you've decided to breed, you need to think about how you will go about it. One bull should be all you need for a start-up, but even that one bull can cost a lot of money (more than a couple of cows), especially considering that you'll be looking for one with good genes to be passed on to your calves. If you start with just one bull – and buy it young enough – your bull should be able to service about 25 cows for about six years.

If you'd rather not spend that much, you could swing for artificial insemination (A.I.). But you'll need a backup bull for those cows, which might not take well to A.I.

A more affordable option could be to share a bull with another outfit so you both can split the cost of buying one. You can also consider a leasing arrangement.

Help

You also want to consider how much help is available to you. If you are starting as a family business, you should have enough hands. Doing it alone is not a good idea!

Hiring additional help will add to your expenses because you must pay them. On the other hand, having to run the entire thing on your own is a lot of work, especially considering that you are not used to the intensity of then work. Weight both options, carefully choosing which works best for you.

If you decide to hire hands, you must think about what aspects of cattle rearing appeal to you the most and then hire out the other aspects. The idea is, if you are passionate about it, it will make doing the work easier and more enjoyable.

It is important to remember that cattle rearing is time and life-consuming as there is a lot of work to be done. You must tend to the cattle while also running a business. So, whether you get help or not, you want to be physically fit and mentally prepared.

People You Need to Know

If you are starting a cattle rearing practice (whether feeder, cow-calf or a combination of both), you will need the contact information of these people:

- A good large animal nutritionist.

- A reliable veterinarian located near you.

- A good extension specialist.

- An experienced cattle-rearer.

- A good butcher or retailer.

21 Tips for Running a Commercially Successful Cattle Rearing Outfit

Whether you're a newcomer to the cattle rearing practice or you've been doing it for years, these tips will help make your practice profitable:

1. When deciding what breed of cattle to rear, consider market trends. You'll want to do a market survey to find out what breeds are in high demand. You can then choose that option or create a niche for yourself if you think you'll be able to get enough customers.

2. More and more people want to buy only grass-fed cattle; consider this option.

3. Try not to skimp on important vaccinations. If you are trying to save costs, there are vaccinations you could do without for many of your cattle, but there are vaccinations that all your cattle must get, and promptly. Prevention is usually less expensive than the treatment or cure. To determine which vaccinations aren't required for your animals and location, you must consult your veterinarian.

4. Don't ever just expect an illness or injury to go away, as things could worsen, and your animal could die. If you notice any of your cattle ill or injured, immediately confer with your veterinarian. Any infirmity greatly reduces the market value of your cattle, and death *cancels it.*

5. If any of your cattle die, know the cause because it could be from something communicable. If you do not know, try to do a necropsy. Once you've been able to ascertain the cause of death, make sure that the rest of your cattle are healthy and safe.

6. Don't just wake up and decide to take your cattle to the market that day without a marketing plan. Developing a marketing plan or strategy involves determining how much your cattle will sell for depending on what the general market price is and the quality of your cattle. It will also include determining the best time to sell.

7. Have a solid network of farmers you do business with. Your network could also include other farmers with whom you could split costs to buy certain equipment or even a bull. Having a solid network can help you get good value for your money.

8. Do everything you can to maintain a good reputation. Cattle rearing exists in a community, and if the people in the community cannot trust you, you won't be able to make much progress.

9. Aside from ensuring that your prices are always legit, patronizing other local businesses for your cattle business will help you maintain a good rep in the community. It could also be your way of contributing to the growth of the community.

10. Do regular analysis and evaluation. It will help you determine what you are doing well and what needs to be improved. It will also help you discover what things you are doing that aren't earning you enough money, which of your cattle isn't bringing in a profit, and which cattle are contributing to the business. Make a plan for the evaluation and decide how often you'll do it.

11. Make sure that you are always up to date with your taxes. Find out if you are eligible for any tax deductions, keeping receipts for every single thing you buy, and invest in a good bookkeeping method; it'll save you a lot of headaches when tax season comes.

12. Make sure that you're always putting effort and money into improving. It is okay if you didn't start with the best equipment and facilities. But as you earn an income from your practice, you'll do well to reinvest and get better equipment and facilities, improve your quality of production and expand your practice if you intend to grow your business.

13. In the same vein, make sure that you are constantly learning about new practices and new technology. Don't experiment with them all, but you'll find something that works for you and gives you good value for your money.

14. Consider livestock insurance. Regardless of how well you plan and how much precautions you take, you might not be able to prevent unfortunate events. Livestock insurance will help cover many of your financial losses caused from cattle accidents, injuries, or illnesses.

15. That you should consult with your vet has been mentioned so often now, but it must also be said that you should be honest with your vet. If they are asking you about your management practices, tell the truth. The truth will help them help you make the right decisions for your practice, at least, as far as your cattle are concerned.

16. Be intentional about your fencing and borders. If you have weak fences and/or porous borders, you could have runaway cattle. Runaway cattle is basically like setting fire to your money unless, by great providence, you can find them.

17. If you are running a feeder practice, consider dehorning your calves. Horns are dangerous and could lead to lots of injuries. And injuries cost money and could reduce their market value.

18. Avoid the craze for modern quick-fixes. Stick with things tested and trusted because cattle cost a lot of money.

19. Buy food in large quantities (as large as a truckful) rather than in small quantities. You'll get better a value for your money when you buy in bulk.

20. Know when it is time to replace your equipment. Old or faulty equipment can become difficult to maintain, and the cost of getting them repaired is ultimately more than the cost of getting a replacement. So, if you've repaired equipment more than twice, you want to think about being financially smart and making a good choice between repairing or replacing the items.

Whatever business you're running, there is a significant possibility that you'll experience a loss. The first time it happens might come as a shock to you, but being prepared for the possibility helps.

Conclusion

Not all the tips and ideas in this book will come to you easily. A few will, while others will take practice. And unfortunately, still others will take trial and error. But try to be patient with yourself. The more time, effort, and proper knowledge you put into your cattle rearing outfit, the better you'll get. This book is not a one-time read. It is a resource you can always come back to as you meet new challenges in your cattle rearing business.

No amount of reading and gathering of information will make you a successful cattle-rearer. You actually have to start something new or change what you're doing to run a successful practice.

It's time to close this book and start carrying out the ideas and suggestions you've read about! Don't forget to come back to it if you ever find yourself in a pickle. And don't forget to involve yourself in the cattle-rearing community. Good luck!

Here's another book by Dion Rosser that you might like

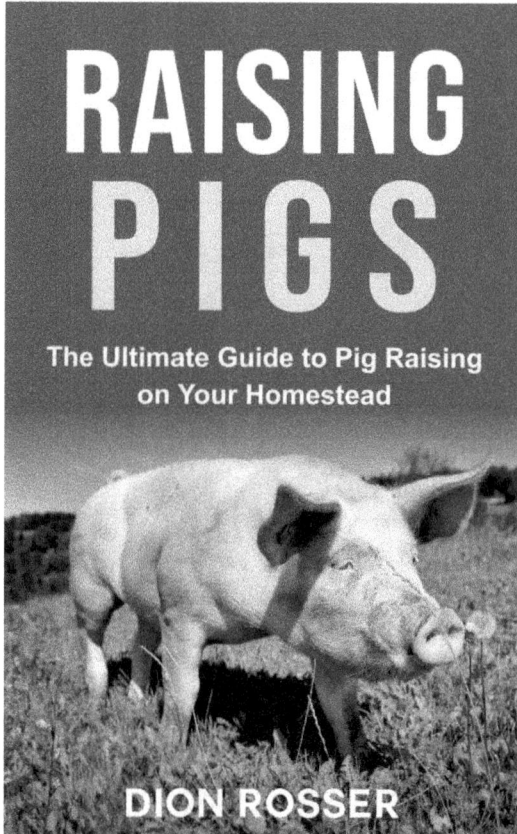

RAISING PIGS

The Ultimate Guide to Pig Raising on Your Homestead

DION ROSSER

References

Amaral-Phillips, D., Scharko, P., Johns, J., & Franklin, S. (n.d.). *Feeding and Managing Baby Calves from Birth to 3 Months of Age.* https://afs.ca.uky.edu/files/feeding_and_managing_baby_calve s_from_birth_to_3_months_of_age.pdf

Beef Cattle Behavior and Handling Understand Behavior to Improve Handling. (n.d.). https://extension.msstate.edu/sites/default/files/publications/pu blications/p2801.pdf

Best Healthy Feed for Beef Cattle. (n.d.). Arrowquip. https://arrowquip.com/blog/animal-science/best-healthy-feed-beef-cattle

Blake, E. 'Skip.' (n.d.). *Cow psychology: Handle them by getting into their heads.* Progressive Dairy. https://www.progressivedairy.com/topics/herd-health/cow-psychology-handle-them-by-getting-into-their-heads

Breeds of beef cattle | Breeds | Beef | Livestock | Agriculture | Agriculture Victoria. (2018). Vic.Gov.Au.

Candi Johns. (n.d.). *6 Reasons to Raise Your Own Meat and How Long it Takes - Farm Fresh For Life Blog - GRIT*

Magazine. Grit. https://www.grit.com/animals/livestock/6-reasons-to-raise-your-own-meat-and-how-long-it-takes-zb0z1601

Cattle Breeds From England. (n.d.). Beef2live.Com. https://beef2live.com/story-cattle-breeds-england-89-106430

Cattle housing equipment. (n.d.). En.Schauer-Agrotronic.Com. https://en.schauer-agrotronic.com/cattle/cattle-housing-systems

Cleaning a Cow: How to Clean a Cow (Beginner's Guide). (2018, March 14). ROYS FARM. https://www.roysfarm.com/cleaning-a-cow/

Cows and heifers calve better later. (2017, May 3). Farm Progress. https://www.farmprogress.com/animal-health/cows-and-heifers-calve-better-later

Gadberry, S., Jennings, J., Ward, H., Beck, P., Kutz, B., & Troxel, T. (2016). *Beef Cattle Production - MP184*. https://www.uaex.edu/publications/pdf/mp184/Chapter3.pdf

Handling Facilities for Beef Cattle. (n.d.). The Beef Site. Retrieved November 5, 2020, from http://www.thebeefsite.com/articles/912/handling-facilities-for-beef-cattle/

https://www.facebook.com/CloverValleyBeef. (2016, September 27). *27 Amazing Facts About Cows That Will Impress Your Friends*. Clover Meadows Beef. http://www.clovermeadowsbeef.com/amazing-facts-about-cows/

Milk Production in Beef Cattle. (2017, March 14). Homestead on the Range. https://homesteadontherange.com/2017/03/14/milk-production-in-beef-cattle/

Niman, N. H. (2014, December 19). Actually, Raising Beef Is Good for the Planet. *Wall Street Journal*.

https://www.wsj.com/articles/actually-raising-beef-is-good-for-the-planet-1419030738

Reproduction in Cattle. (2019). Peda.Net. https://peda.net/kenya/css/subjects/agriculture/form-3/lsab/ric

Seven Fencing Tips for Cattle | Ag Industry News - Farm and Livestock Directory. (n.d.). Farmandlivestockdirectory.Com. https://farmandlivestockdirectory.com/seven-fencing-tips-for-cattle/

Steer Vs. Bull. (n.d.). Animals.Mom.Com. https://animals.mom.com/steer-vs-bull-3150.html

Top Ten Considerations for Small-scale Beef Production. (n.d.). Small Farm Sustainability. https://www.extension.iastate.edu/smallfarms/top-ten-considerations-small-scale-beef-production

www.ingramcontent.com/pod-product-compliance
Lightning Source LLC
Chambersburg PA
CBHW050644190326
41458CB00008B/2409